SOCIETY FOR EXPERIMENTAL BIOLOGY

seminar series: 22

INSTRUMENTATION FOR ENVIRONMENTAL PHYSIOLOGY

Instrumentation for environmental physiology

Edited by

B. MARSHALL

Scottish Crop Research Institute, Dundee

F. I. WOODWARD

Department of Botany, University of Cambridge

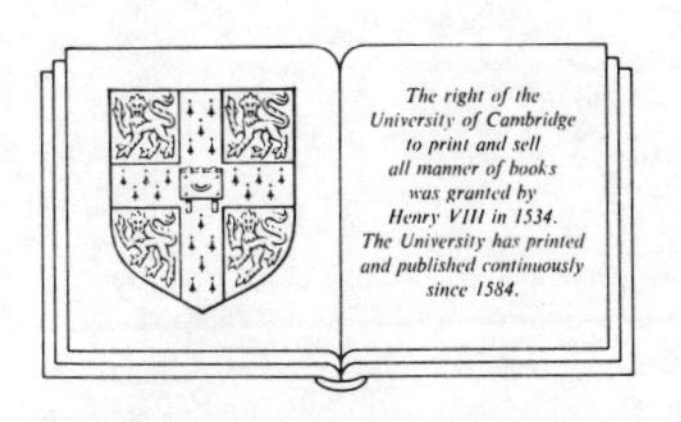

CAMBRIDGE UNIVERSITY PRESS

Cambridge

London New York New Rochelle

Melbourne Sydney

Published by the Press Syndicate of the University of Cambridge
The Pitt Building, Trumpington Street, Cambridge CB2 1RP
32 East 57th Street, New York, 10022, USA
10 Stamford Road, Oakleigh, Melbourne 3166, Australia

First published 1985

Printed in Great Britain at the University Press, Cambridge

British Library cataloguing in publication data

Instrumentation for environmental physiology. – (Society for Experimental Biology series; 22)

1. Physiology – Measurement 2. Ecology
I. Marshall, B. II. Woodward, F.I. III. Series
574.1 QP82

Library of Congress cataloging data available

ISBN 0521 25399 3

Contents

Preface

Advances in environmental physiology are heavily dependent on developments in associated instrumentation. This very strong tie ensures a healthy interest in instrumentation and a concern for appropriate development. The evolution of instrumentation is a continuous process, however times must be selected for 'state-of-the-art' reviews of technology. The last major review on instrumentation for plant ecophysiology was by Sestak, Catsky and Jarvis in 1971 (Plant Photosynthetic Production: A Manual of Methods, Dr W. Junk, The Hague). The date of this publication has been taken as the starting point for the reviews comprising this book.

The emphasis on instrumentation for studies on convenient, sedentary plants still remains, although devices may be equally applicable to animal ecophysiology.

This volume aims to review significant developments in instrumentation from 1971 and covers a wider range of measurements than previously. The book is effectively divided into two parts. The first is concerned with the measurement of major environmental variables, with a general applicability to biological organisms. The second part is concerned with physiological measurements on plants. In presenting their reviews, the authors were asked to consider several points: relevant biological ranges of measurement; terminology and units; accuracy, precision, response times and long term stability; techniques of calibration and provision of standards; practical problems and data handling. It was not possible within the scale of this book to discuss every aspect of environmental measurement. In particular, the edaphic environment has not been dealt with in any detail and this important topic may well form the subject of future publications.

This volume is a publication of review papers presented at the Annual Conference of the Environmental Physiology Group (Society for Experimental Biology) and held jointly with the Association of Applied Biologists and the British Ecological Society, at the University of Hull, from the 6th to the 8th April 1983.

We are grateful to all the contributors for their carefully presented manuscripts, to Dr L.D. Incoll (University of Leeds) and Dr Anne Brown (University of Hull) for assistance with organising the Conference, and to Mr P. Smith (Scottish Crop Research Institute) for painstakingly proof-reading the final manuscripts.

B. Marshall
F.I. Woodward

Chapter 1: Measurement – A game of Snakes and Ladders

J. L. Monteith
Department of Physiology and Environmental Science,
University of Nottingham School of Agriculture, Sutton Bonnington,
Loughborough, Leicestershire LE12 5RD.

Most of us, when young, played Snakes and Ladders before progressing to more sophisticated games like Monopoly, based on the same principle but demanding an element of skill for success as well as sheer luck in throwing the dice. Now we are engaged in Research, another game which needs both skill and luck. The analogy of Research as an extended, and sometimes very frustrating game, seems particularly relevant to the subject of this meeting.

Consider the structure of the Snakes and Ladders board. At the bottom of the board, starters are encouraged to find more ladders than snakes so initial progress is relatively fast. Near the top, snakes predominate, some of fearsome length. So a plot of the *average* progress of a player against time would be something like the rectangular hyperbola often used to represent the relation between the photosynthetic rate of a leaf and its irradiance. But the performance of an *individual* player would show very large deviations both above and below the average response. This individual response, up ladders and down snakes, is very like the progress of a research project – with one major difference. Even when a project is wound up we rarely feel we have 'won' in the sense of finishing a game. Instead there are various stages of success. Preliminary results are discussed with colleagues over coffee. Then there is enough material for an internal seminar or even for a meeting of the Society for Experimental Biology. Eventually, the story becomes plausible enough to satisfy the referees of a scientific journal.

Many of the snakes and ladders of research refer to instrumentation. Ladder: head of department finds source of funds for new gas analyser. Snake: head of department turns wrong valve and injects contents of manometer into analyser. Ladder: ideal weather for first complete set of observations. Snake: cables gnawed through by vermin during the night. Looking back over 30 years of research, I can identify a number of snakes and ladders in the development of instruments for environmental physiology. The most obvious ladder has been the improvement of methods for recording. For my Ph.D. (on dew), I had to sit up all night measuring temperature profiles with a mirror galvanometer and a lamp and scale, equipment common in physics laboratories a century ago. In the 1950s, Kent recorders developed for industrial use were very reliable for measurement with resistance thermometers but this type of instrument could not be used with thermocouples until stable DC amplifiers came on the market in the late 60s. At about the same time, we were beginning to use data loggers both for analogue and digital signals but the first system I saw, at Davis, California, produced in one day enough punched cards to fill a cabin trunk. Then came paper tape, magnetic tape, and finally solid-state memory devices – a very long and steep ladder of progress.

Some of the associated snakes have been clearly identified already. With contemporary systems of data collection, there is a strong temptation to amass too many records and indigestion soon develops. This is soon followed by constipation - the inability to get rid of waste material. More and more people haven't time to write the papers that fewer and fewer people have time to read. A more subtle snake is the loss of direct contact between the experimenter and the environment he is measuring or the organism responding to that environment. Often it is the unexpected phenomenon, detected and explored, that advances science. I worry that if recording systems become too sophisticated, we may sometimes miss significant signals because a computer program has rejected them as "noise".

Another major improvement has been the reliabiliy and precision of electronic equipment for detecting and amplifying small signals. The infra-red gas analyser is a conspicuous example. In 1958, I paid my first visit to the Infra-red Development Company at Welwyn Garden City to discuss with W. B. Bartley the possibility of obtaining a CO_2 analyser with a full-scale deflection of $\pm$ 5 ppm which we wanted for profile work. This was an order of magnitude more sensitive than anything on the market at the time. To their great credit, IRDC did make an instrument to this specification but the detector had to be refilled and resealed every few weeks. Nowadays detectors run for years and solid-state circuits have made IRGAs so portable that they can be operated by one man in the field.

The snake associated with better reliability, accuracy, and precision is another subtle one. There is a danger of becoming obsessed with the quest for perfection in one class of instrument when the success of a research project is limited by the performance of some other component of the system. How can we compare the performance of different instruments to minimise the danger of being bitten by this snake? We can appeal to a principle which many people are aware of intuitively but which is rarely applied with rigour in environmental physiology – the admirable principle of Minimum Work.

Suppose we are trying to determine a quantity Y which is a product of measured variables x_1, x_2, x_3, etc., with associated fractional errors of $\triangle_1$, $\triangle_2$, $\triangle_3$, etc. We can postulate that the amount of work needed to keep an error below $\triangle$ is proportional to $1/\triangle$. Then if the error in Y is to be kept below some chosen value, it can be shown that the total work is minimised when all the errors are the same. It can further be shown that if any one fractional error is less than about 40% of the mean error, it can be ignored. So the principle of Minimum Work tells us to keep fractional errors similar and to stop fussing about an instrument giving errors less than 40% of the mean error of other instruments.

This book is concerned with instrumentation of the last decade but in identifying major ladders in research, I would like to go back a little further. In fact, the 50s and 60s probably produced more new types of instruments than the 70s have done; the main advances in the past 10 to 15 years have been in performance and portability.

Measurements of radiation exchange at the surface of vegetation or of an animal became possible with the development of net radiometers. The original Gier and Dunkel type with a ventilated plate has been almost entirely superseded by the Funk type with a polyethylene dome. The measurement of radiation interception by vegetation was facilitated by the development of linear thermopiles enclosed in glass tubes, a design first developed by Isobe in Japan.

Studies of plant water relations in the field have made rapid progress because of the availability of instruments such as the diffusion porometer, the thermocouple psychrometer, the pressure bomb, and the neutron probe.

All these instruments have been greatly improved since prototypes first appeared and are now stalwart ladders on the board of environmental physiology. All were developed to meet a need. The corresponding snake is the kind of technique which goes looking for a problem to get its fangs into. Some aspects of "remote sensing" seem to be in this category at the moment but this rapidly expanding arm of technology contains new ladders which the ecologists must learn to climb. Many of the ladders they use now involve remote sensing in one form or another.

What instruments need to be developed in the next 10 to 20 years? I believe the principle of Minimum Work should be applied in a very broad way to restrain us from improving devices which are already working well and to encourage us to think about bits of the environment and responses of organisms we have neglected because convenient sensors were not available. The abstracts for this meeting suggest that we shall hear little about measurements in the soil. How can we measure soil water content in layers a few centimetres deep instead of averaging over 20 or 30 centimetres with a neutron probe? How can we measure fluxes of nutrient ions in the soil? Moving to organisms, is there any hope of being able to measure leaf turgor directly rather than in the inaccurate difference between two large potentials? How can we measure the metabolic rate and the loss of heat by convection and evaporation from a free-ranging animal? In the aquatic environment, there are probably equivalent needs which are restricting the progress of research.

The abstracts also suggest that we shall hear little about instruments for measuring the chemical components of the atmosphere other than CO_2 and H_2O but this is already a major area of research of special interest to the British Ecological Society and one which will continue to expand over the next decade with incentives to develop similar techniques for monitoring concentrations and fluxes of elements and compounds within ecosystems. Other major areas for instrumental developments of special interest to the Association of Applied Biologists are aerobiology and soil biology. In the atmosphere, radar and other forms of remote sensing will be used increasingly to monitor the distribution and movement of insect pests. I do not know of any comparable attempts to monitor the behaviour of invertebrates within the soil matrix, a formidable but hopefully not an insuperable problem for instrumentation.

I wish you both skill and luck in your search for the ladders which will help you to explore aspects of the environment and physiological responses which we cannot yet measure. Try to avoid the more obvious snakes, but don't be discouraged when you encounter one. Research – and life – would be rather dull without occasional backward moves to make us work harder and more efficiently!

Chapter 2: Radiation

J. E. Sheehy
Grassland Research Institute, Hurley, Maidenhead,
Berkshire, SL6 5LR, England.

The exceptional place held by light in science might be thought to stem simply from our possession of sight. However, its scientific importance is far greater than sound which can also be perceived by one of our senses. The behaviour of light has revealed much about the structure of matter, furthermore, it is part of an extensive class of phenomena known as radiations. The main objective of this chapter is to discuss the nature of radiation, the terminology used in its description and some of the instruments used for its detection. A secondary objective is to describe some novel developments and their potential in agricultural research.

The nature of radiation

Newton described light as a stream of particles, but his corpuscular theory was abandoned when it failed to explain interference. Most scientists are aware of the simple experiment of Dr Young in 1801. Referring to Fig. 2.1a, when light passes through a pinhole in an opaque screen *AB* to another screen *CD*, some amount of light will fall on all points between *X* and *Y*. If another pinhole is made in the screen an additional amount of light will fall between *X* and *Y*. According to Newton's corpuscular theory the amount of light at all points between *X* and *Y* would result from the addition of the new and original amounts. However, it is observed that there are alternating bands of light and darkness, a photoelectric cell can be used to measure the amount and distribution of the light. A simple explanation of this observation is provided by considering light to be composed of waves rather than particles. If the pathlengths between the pinholes and the screen is such that a trough of one wave and a peak of another coincide, there is destructive interference and darkness (Fig. 2.1b). Where the two apparent sources reinforce each other there are bright bands (Fig. 2.1c).

Maxwell derived the equations of the electromagnetic theory which described radiation, including light, in terms of combined electric and magnetic disturbances transmitted as waves, The wave theory has been satisfactory for describing the reflection, transmission, diffraction and refraction of radiation. At the long wavelengths of radiowaves it has also been satisfactory for describing the interaction of radiant energy and matter, such as the effects of the radiation on receiving aerials. However, the observed wavelength distribution of radiation emitted by a hot body could not be described at short wavelengths using the wave theory. At the wavelengths of light, towards the violet end of the spectrum, the wave theory failed to describe correctly the relationship between radiation and matter.

The wave theory was unable to provide a satisfactory explanation of why emission

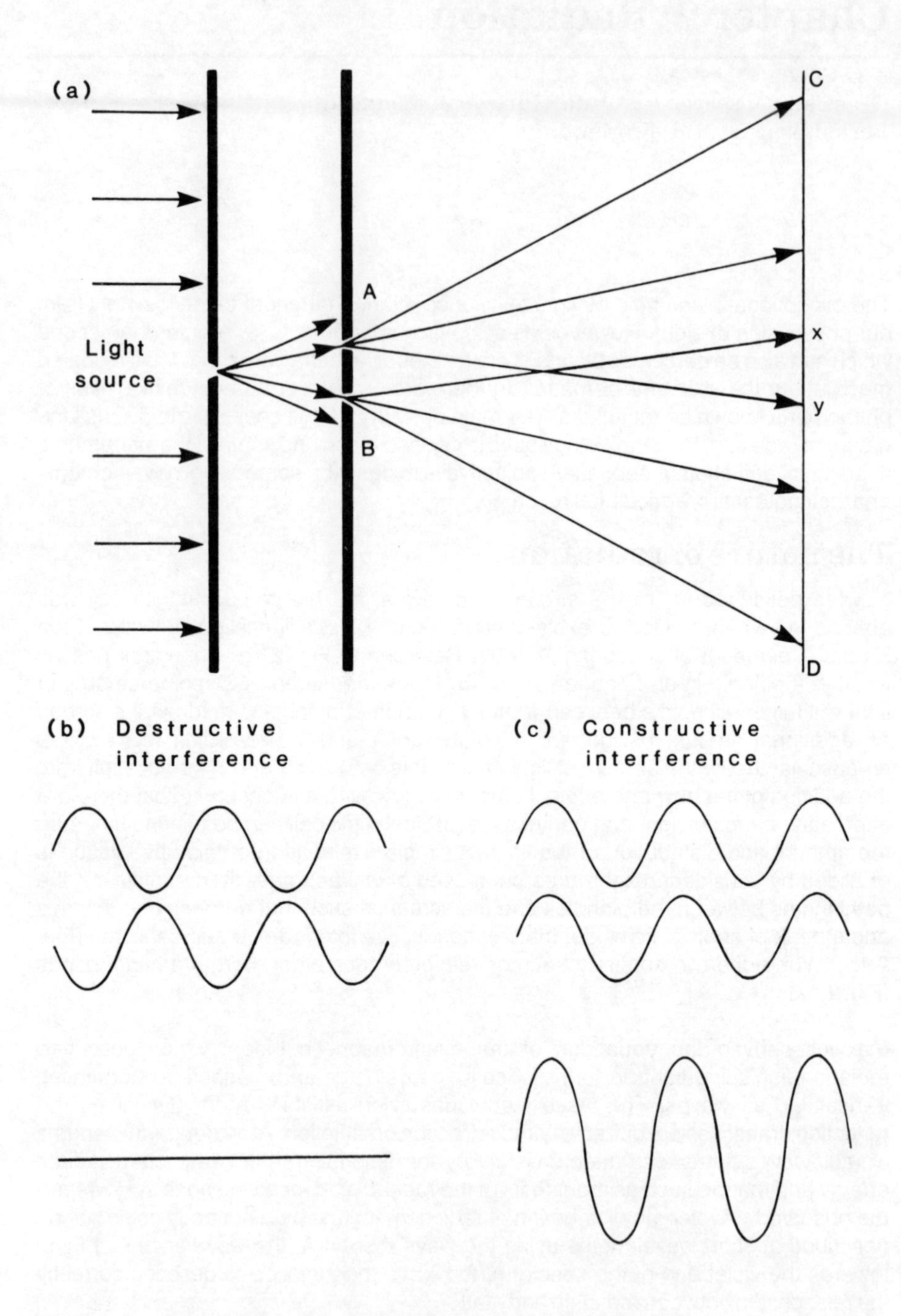

Figure 2.1 (a) Young's pinhole experiment and the interference of (b) destructive and (c) constructive interference.

of electrons from matter exposed to light depended on the quality and not the quantity of light. Below a certain frequency electrons are not emitted. In 1900 Planck introduced the idea that the emission of radiation from matter could take place only in discrete amounts and explained successfully the relation. Einstein, in 1905, suggested that when light gave up its energy to matter, it behaved as if it was composed of discrete amounts. These were called quanta, each quantum having an energy $h\nu$, where h is Planck's constant and ν is the frequency. The concept of a lower frequency limit is common to all interactions of radiation with matter with the exception of inducing a current in a conductor. For example, in the case of radio waves it is the presence of free electrons in the conductor which permits the non-quantized behaviour of the waves.

A quantum of radiation is a small packet of pure energy. When it is absorbed by matter,because it has no mass, it can therefore be considered to have ceased to exist. Because of this property it plays a useful role in the conservation of energy.

Thus, the ability of matter and energy in the form of radiation to interchange characteristics and to interact is a cornerstone in our understanding of the universe. It has been experimentation into the dual nature of light waves and quanta that has led to this deeper understanding of the structure of matter. Nevertheless, the apparent duality of light and the difference between the theories which describe it lie at the heart of most of the difficulties we encounter in the terminology of its description.

Terminology and units

The multi-disciplinary nature of biological research has introduced a wide range of terminology. Often a term is as common in one discipline as it is uncommon in another and occasionally it can have a completely different meaning. It is not surprising therefore that misunderstandings occur across discipline boundaries. Many definitions are sonorous and have great appeal as poetry, but little as a form of precise communication. Terms such as "photosynthetic photon flux fluence rate (PPFFR)" and "photosynthetically active photon flosan (PAPF)" have a nice 'ring' to them, but are somewhat obscure. Equally confusing is the use of the term density in "radiant flux density" to convey the idea of something with the dimensions of an area rather than a volume. An excellent analysis and synthesis of terminology associated with light measurements has been presented by Bell and Rose (1981) and is also discussed by Incoll (Chapter 13). Table 2.1 shows some of the terms used to describe the radiation environment in recent editions of three major journals. It is evident that there is no uniform terminology.

The relation between a quantum and its energy content is a function of wavelength. Thus for monochromatic radiation changing between the terminologies of energy and quantum numbers presents no great difficulty. Problems occur when there are a number of wavelengths and the amount of radiation at each wavelength is unknown. Nevertheless, it is easier to visualise radiation in terms of quanta than in terms of waves. Thus, we shall first examine terminology using quantum concepts, although it must be realised that it was originally derived to describe the movement of electromagnetic waves and their interaction with matter. The path which a quantum describes in space is a flux line; a number of quanta moving through space in unit time

Table 2.1 Terms used to describe radiation environment in recent editions of three major journals.

Term	Units	Journal
Photon flux density	mol m^{-2} s^{-1}	J. exp. Bot.
Illuminated to	μE m^{-2} s^{-1}	J. exp. Bot.
PAR	W m^{-2}	J. exp. Bot.
Photon flux density in the photosynthetically active range	μmol m^{-2} s^{-1}	Plant Cell & Environ.
High fluence rate	W m^{-2}	Plant Cell & Environ.
Total photon irradiance	μmol m^{-2} s^{-1}	Plant Cell & Environ.
Photon fluence rate	μmol m^{-2} s^{-1}	Plant Cell & Environ.
High intensity light	no units	Pl. Physiol.
Light intensity	μE m^{-2} s^{-1}	Pl. Physiol.
Illumination	W	Pl. Physiol.
Irradiance energy	μE m^{-2} s^{-1}	Pl. Physiol.
Intensity was increased to full irradiance	μE m^{-2} s^{-1}	Pl. Physiol.

comprises a flux. A number of quanta incident on unit area of a plane surface in unit time comprises an irradiance. The net number of quanta passing through unit area at right angles to their direction in unit time comprises a flux area density. The term flux density is one of those terms which has different meanings in different disciplines and unless essential it is probably best avoided.

To describe the amount of radiation emitted by a radiation source we have to count the number of quanta leaving the source. If the source is small and we imagine it to be surrounded by an imaginary spherical shell, the number of quanta passing through the shell in unit time gives a measure of the radiant strength of the source. A sphere subtends a solid angle of 4π and so the source strength can be defined as the number of quanta emitted into the solid angle 4π in unit time. The definition of intensity is therefore: the number of quanta emitted into unit solid angle in unit time. For a large source another way of assessing its strength is to consider the number of quanta leaving unit area of its surface in unit time – this is termed the excitance. A large source can be viewed from many different angles and so it is useful to describe the apparent strength of the surface from its viewing angle. The apparent surface is the projection

Table 2.2 Photons: quantities, symbols, units and recommended terms.

Quantity	Symbol	SI unit	Recommended term (Bell and Rose, 1980)
Number of photons	N	mol	Number of photons
Number of photons passing through some region of space in unit time	$\phi_p = \frac{dN}{dt}$	mol s^{-1}	Photon flux
The net number of photons passing through unit area perpendicular to their direction in unit time	$\frac{d\phi_p}{dA}$	mol m^{-2} s^{-1}	Photon flux area density
Number of photons emitted into unit solid angle in unit time	$\frac{d\phi_p}{d\Omega} = I_p$	mol s^{-1} sr^{-1}	Photon intensity
Number of photons passing through unit area of a surface in the direction of emission per unit solid angle per unit time	$\frac{1}{\cos\theta}\frac{dI_p}{dA}$	mol m^{-2} s^{-1} sr^{-1}	Photon radiance
The number of photons incident on a surface per unit area per unit time	$\frac{d\phi_p}{dA}$	mol m^{-2} s^{-1}	Photon irradiance
Number of photons emitted by a surface per unit area per unit time	$\frac{d\phi_p}{dA}$	mol m^{-2} s^{-1}	Photon excitance

of the emitting surface at right angles to the line of view. Thus the number of quanta emitted per unit area of the apparent surface per unit solid angle is known as the radiance. It should be noted that strictly a photon is a quantum of visible radiation, but it is often used to describe any quantum of radiation.

The same concepts are used to describe radiation in energy terms, the difference between the terminologies being denoted by the prefixes radiant, quantum or photon. Tables 2.2 and 2.3 give summaries of the quantities, units and terminology likely to be of most use to the environmental physiologist.

Incoll, Long and Ashmore (1977) drew attention to the two definitions of an einstein now in use. In photochemistry an einstein is the quantity of energy in Avogadro's number of photons; the energy of an einstein according to this definition varies with wavelength. In plant physiology (Shibles, 1976) an einstein is used to describe Avogadro's number of photons ($6.023\ 10^{23}$ photons i.e. a mole of photons). The term einstein is not part of the SI system and should not be used. Photon or quantum irradiance should have the units mol m^{-2} s^{-1}, with the waveband or wavelength also defined. In an extremely elegant and relevant piece of work, McCree (1972) showed that for an individual leaf, quantum irradiance in the waveband 400-700 nm was close enough to the "ideal" measure of photosynthetically active radiation for all practical purposes.

Table 2.3 Radiation: quantities, symbols and recommended terms.

Quantity	Symbol	SI unit	Recommended term
Energy in the form of electromagnetic radiation	Q_e	J	radiant energy
Rate of propogation of radiant energy	$\phi_e = \frac{dQ}{dt}$	$J\,s^{-1} = W$	radiant flux
Net radiant flux through unit area normal to the area	$\frac{d\phi_e}{dA}$	$J\,m^{-2}\,s^{-1} = W\,m^{-2}$	radiant flux area density
Radiant flux emitted into a unit solid angle	$\frac{d\phi_e}{d\Omega} = I_e$	$J\,s^{-1}\,sr^{-1} = W\,sr^{-1}$	radiant intensity
Radiant flux passing through unit area of a surface in the direction of emission per unit solid angle	$\frac{1}{\cos\theta}\frac{dI_e}{dA}$	$J\,m^{-2}\,s^{-1}\,sr^{-1} = W\,m^{-2}\,sr^{-1}$	radiance
Radiant flux incident on unit area	$\frac{d\phi_e}{dA}$	$J\,m^{-2}\,s^{-1} = W\,m^{-2}$	irradiance
Radiant flux emitted per unit area	$\frac{d\phi_e}{dA}$	$J\,m^{-2}\,s^{-1} = W\,m^{-2}$	radiant excitance

Type of radiation detector

There are two basic types of radiation detector. The first type depends on the transformation of absorbed radiation into heat and it is this thermal effect which is measured. Such detectors are known as thermal detectors. The second category depends on the quantum activated release of electrons which participate in current flow. Such detectors are called photo-detectors. The essential difference between the two types of detector is that in a thermal detector the energy of absorbed quanta, some of which release bound electrons, is dissipated throughout the molecular system as heat. The thermal detector depends on the total energy of the absorbed radiation. In contrast, only the incident quanta which are effective in releasing bound electrons are detected in the photo-detector. The photo-detector depends for its action on the number of such quanta only. Other absorbed quanta are dissipated as heat, but the thermal capacity of the detectors is usually large enough for this not to be a serious problem.

The absorbing surface of a thermal detector is usually blackened; consequently, they have a much wider spectral response than the photo-detectors. It is probable that most thermal detectors have a uniform spectral response in the 400-700 nm waveband, although in the infra-red waveband this may not be so (Betts, 1965). Thermal detectors have response times of the order of seconds, at best milliseconds, compared with microseconds for photo-detectors. The physical properties of the different types of radiation detector are fully discussed by Smith, Jones and Chasmar (1968) and Kubin

(1971). Devices such as ceramic pyroelectric, ozalid paper, thermomechanical and distillation radiation detectors are less common and will not be discussed.

There are three major types of thermal detector: the thermopile, the bolometer and the pneumatic cell. In the thermopile detectors one set of thermocouple junctions is connected to a blackened surface which is exposed to the radiation and the other set of junctions is connected to a white surface which is exposed or a heat sink which is unexposed to the radiation. The radiant energy absorbed by the blackened surface is transformed into heat, less energy is absorbed by the white surface, and the temperature difference between the two sets of junctions is proportional to the incident radiation. The thermopile is constructed so that ambient temperatures affect both sets of junctions equally and therefore have no effect. Due to intrinsic noise problems the minimum power detectable with the thermopile devices is approximately 10^{-10} W. Many of the radiometers used in environmental physiology are thermopile types (Kubin, 1971; Anderson, 1971; Fritschen and Gay, 1979).

The bolometer depends on the change in resistance of a piece of blackened material exposed to radiation often relative to a similar piece which is unexposed. Platinum is commonly used in bolometers because it can be made into thin strong strips. At the temperature of liquid helium bolometers can detect 10^{-12} W.

The Golay cell is a pneumatic detector in which a gas filled chamber is heated via contact with a blackened radiation absorbing surface. The gas expands and this can be detected either optically by the movement of a flexible mirror or electrically using a bridge circuit. The gas chamber is arranged in a differential manner so that changes in ambient temperature have no effect. Their sensitivity is of the order of 10^{-11} W.

There are three common types of photo-detector: the photo-conductive, the photo-emissive and the photovoltaic. The photo-conductive or resistive type depends on the release of electrons within the radiation sensitive material upon exposure to radiation. The presence of the extra electrons lowers the electrical resistance because they can participate in the current flow. These types of device are usually very temperature sensitive. In the dark their resistance is of the order of MΩ and in the light it can be of the order of 100 Ω; the maximum power dissipation is approximately 500 mW.

Photo-emissive detectors, photomultipliers, are used to measure low levels of radiation. Quanta are absorbed by a suitable material and electrons are released, these are focussed and accelerated towards another surface. Upon impact, the electrons are sufficiently energetic to release further electrons in a process called secondary emission. These secondary electrons are subjected to the same process and after several stages a substantial gain in electrons is achieved.

The most commonly used photo-detector is the photovoltaic detector. The detector is based on a PN semiconductor junction (Chappell, 1976). When a suitable quantum is absorbed a hole electron pair is formed. If they are produced away from the interface between the P and N materials they quickly recombine; however, if they are produced near the junction they separate. Moved by forces of attraction they can form a current in a conductor. If the load resistance is small the current flow is linearly proportional to the number of quanta absorbed. More detailed descriptions of optoelectronic circuits can be found in Chappell (1976). It is important to note that manufacturers provide calibrations for photo-detectors in energy units and not in quantum units.

An understanding of the fundamental difference between thermal and quantum detectors is of importance when measuring radiation in spectrally selective media. The results produced by the two types of instrument can be quite different, even when they both have constant responses at all wavelengths. This is best illustrated by a simple example.

Example

Four micromoles of quanta are incident on unit area of the non-reflecting surface of a spectrally selective medium, per unit of time. The radiation is composed of two wavelengths, λ and 2λ, and there are equal numbers of quanta at each wavelength. The radiation is to be measured at some position, A, where 90% of the quanta at λ have been absorbed and 50% at 2λ; using a thermal detector which produces a constant output for equal energy inputs at all wavelengths and a quantum detector which produces a constant output when exposed to equal numbers of quanta at each wavelength. For the purpose of this example the detectors will be assumed to have perfect sensitivity. Estimate the percentage transmission of radiation through the medium to the measuring position in energy terms and quantum terms.

First consider the measurement made using the quantum detector:

the total quantum irradiance $= 4\ \mu\text{mol m}^{-2}\text{ s}^{-1}$,
the total quantum irradiance at A $= 0.1(2) + 0.5(2)$,
$= 1.2\ \mu\text{mol m}^{-2}\text{ s}^{-1}$.
The percentage quantum transmission $= 30\%$.

Next consider the measurements made with the thermal detector:
if we assume that the energy of a quantum at λ is E, the energy at 2λ is $0.5E$.

The total irradiance $= 2(E) + 2(0.5E)$,
$= 3E\text{ W m}^{-2}$,
the total irradiance at A $= 0.1(2(E)) + 0.5(2(0.5E))$,
$= 0.7\text{E W m}^{-2}$.
The percentage energy transmission $= 23.3\%$.

If the attenuation coefficients are reversed the percentage energy transmission is 36.7%.

This example illustrates quite clearly the difference between the detectors and illustrates the importance of deciding whether it is an energy or a quantum process which is being studied.

Common assumptions and radiation measurements

Only absorbed radiation can drive photochemical processes. However, the two quantities most frequently measured are incident and intercepted irradiance. The radiation balance of a point (P) with Cartesian coordinates (x,y,z) is shown in Fig. 2.2a; consider a small rectangular body *ABCDEFGH* (Fig. 2.2b) centered at the point (P) and let I denote monochromatic irradiance.

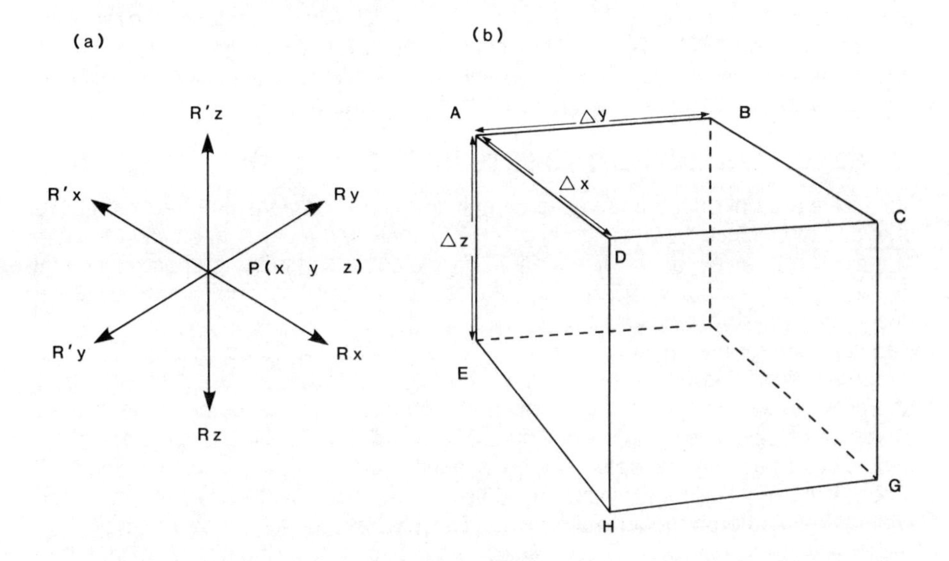

Figure 2.2 (a) The radiation balance of a point P, (b) a small rectangular body in a monochromatic radiation environment.

The absorbed radiation, I_{ab}, is equal to the sum of the radiations incident on all surfaces less the sum of the radiations leaving each face. The incident irradiance, which is commonly measured, is the downward flux of radiation, I_z, incident on face *ABCD*. I_z will only approximate to I_{ab} when a canopy completely covers the ground; a negligible amount of radiation reaches the soil and there is little radiation reflected from the crop surface. When the canopy is incomplete (row structure or isolated plants) then I_z is a poor approximation of I_{ab}.

Light interception is usually defined as that fraction of the light incident on the top of a canopy which fails to reach the ground. Referring to Fig. 2.2b, the fraction intercepted

$$F = 1.0 - (I_{z+\Delta z}/I_z) \qquad 2.1$$

If we assume that the rectangular body is part of a continuous, homogeneous medium in the x and y directions then there will be no net exchange of radiation in those directions. If the radiations reflected from the canopy surface and the ground are negligible then

$$FI_z = I_{ab} \qquad 2.2$$

If the canopy is incomplete there will be significant exchanges of radiation in the x and *y* directions. However, the approximation can still be valid if averaged over a day or more, as the exchanges of radiation in the *xy* plane will tend to average to zero.

The overall conclusion from this section is that for individual plants, spaced plants, small swards or crops with high reflectance, care must be taken to define the radiation environment and the implications for measurement. It is interesting to note that an apparatus which measures radiation absorption and photosynthetic quantum yields for whole plants is described by Oquist, Hallgren and Brunes (1978).

Instrument developments (1971-1983)

It is in the area of quantum detection that instruments for measuring radiation have changed most since 1971. Biggs *et al.* (1971) published a paper describing the use of a silicon photocell, optical filters and a cosine correcting head to effectively count quanta in the waveband 400-700 nm. Variations on their basic design are common and widely used to measure "photosynthetically active radiation". Woodward and Yaqub (1979) further developed the concept when they combined an integrator, and temperature measuring feature, with the quantum detector. More significantly they attempted to "match" the relative spectral response of the detector to the relative quantum response of photosynthesis. Woodward (1983) further developed the concept of "matching" the detector response to the quantum response of a biological process in an instrument which, in addition to a "photosynthetic" response, had various "photomorphogenic" responses. The precise consequence of "matching" responses on the measurement of radiation are discussed in the next section. Fitter, Knapp and Warren Wilson (1980) took a somewhat different approach and used a similar system to measure the radiant energy in the waveband 400-700 nm.

Hiroto, Takeda and Saito (1975) mounted fifteen photodiodes in a linear array, each having an appropriate optical filter and a cosine correcting head, which was used to measure the transmission of photosynthetically active radiation in rice canopies. Williams and Austin (1977) described a similar arrangement incorporating a group of four photocells which remained above the canopy; the circuitry was arranged to provide a digital display of per cent transmission. Despite the absence of a "flat response" in the 400-700 nm waveband the per cent transmission was similar to that measured using tube solarimeters. Another similar device which used selenium cells was described by Muchow and Kerven (1977), unfortunately they did not provide a detailed spectral response or evaluation of the instrument's performance within a canopy.

New methods of constructing tube solarimeters have been described by Constantine (1979) and Klabzuba (1975). Luxmoore, Millington and Aston (1971) described modifications to tube solarimeters to enable them to measure energy on the upper and lower surfaces, thus providing additive or net measurements. They also described the use of a filter for separating the visible and shortwave infra-red wavebands. Baille, Guicherd and Mermier (1978) have described a hemispherical solarimeter for measuring the radiation in the ultra-violet, visible and shortwave infra-red wavebands. Palmer (1980) used the data of Szeicz (1974) to help provide new estimates for the proportion of near infra-red radiation to be used in the calculations of visible irradiance, when solarimeters are used to separate shortwave irradiance into the visible and infra-red wavebands. He suggested that for spot measurements made with Wratten 88A filters the fractional transmission can be calculated using the equation

$$I_v/I_v(0) = (I-0.44I_r)/(I(0)-0.44I_r(0)), \qquad 2.3$$

where $I_v(0)$, $I(0)$, $I_r(0)$ are visible, total and shortwave infra-red irradiances above the canopy and I_v, I, I_r below the canopy.

The technique of hemispherical photography was reviewed by Anderson (1971). Steven (1977), Steven and Unsworth (1980) and Woodward and Sheehy (1983) described the radiance of the sky. Measurements by Grace (1971) suggested that the concept of a "standard overcast sky" provided a reasonable description of the observed mean pattern in Britain. Hooper (1976) used the technique in the short dense canopies of fully developed wheat and barley stands to provide a good description of gap frequency and per cent light transmission. Bonhomme and Chartier (1972) described a device to facilitate the rapid interpretation of hemispherical canopy photographs. They projected the image and rotated an array of photocells beneath the image; by recording position and irradiance they could produce a mapping of the gap-area distribution.

Eckardt, Methy and Sauvezon (1969) constructed a device which could measure the actual distribution of irradiance in several selected wavebands within the canopy. Sinclair and Lemon (1974) used a similar concept to measure the actual distribution of photosynthetically active radiation within a corn crop in clear and overcast conditions. They used a quantum sensor with a uniform sensitivity in the waveband 400-700 nm, a response time of 100 μs, a deviation of less than +2% from the ideal cosine response over the angle 0-80°, a diameter of 50 mm and a height of 75 mm. Sheehy and Chapas (1976) used a similar technique to examine the irradiance distribution within grass canopies, in clear and overcast conditions, using a small sensor with an active area of 2 x 1 mm. When measuring the actual distribution of irradiance within a canopy the influence of sensor size on the observed distribution has to be considered (Woodward and Sheehy, 1983).

The basic thermopile method of measuring net radiation, the difference between the downward and upward irradiances of total shortwave and longwave radiation, has not changed since the review of Kubin (1971). Campbell, Mugas and King (1978) showed that a Kipp solarimeter could be modified to give a good measure of the shortwave and longwave irradiance by changing the glass domes for a single polythene dome. Fritschen and Gay (1979) have provided a useful review of commercially available devices and Woodward and Sheehy (1983) have described some of their design characteristics and applications. When measuring the exchange of radiant energy in the natural environment, interference due to the structure supporting the radiometers must be negligible, a problem considered by Reyenga and Dunin (1975).

Spectroradiometric techniques have benefited from the introduction of the microprocessor and it is now possible to scan the waveband 300-1100 nm in approximately 30 s and produce a record in energy or quantum units, as for example with the LI-COR LI 1800.

Developments in electronics during the past decade have meant that low cost, low power consumption integrators could be constructed and such devices have been described by Campbell (1974) and Burgess and Cox (1975). The interpretation of integrator results has always been difficult because of the non linear response of plants to irradiance. Woodward and Yaqub (1979) provided a novel solution by constructing integrators which had threshold responses. The use of a number of integrators with different thresholds can be used to ameliorate the problem of non linearity of responses.

The problems of light sources such as power output and spectral composition are dealt with elsewhere. Developments in the area of reflectance measurements and remote sensing are desribed by Biscoe and Jaggard (Chapter 12), the notable review of Bunnik (1978) should also be of assistance in this area.

Detector and plant responses

It is possible to shape the spectral responses of a detector to match the spectral response of the plant process. Woodward (1983) described an instrument which could be used for measuring "matched" radiation in several wavebands. In one waveband the spectral response closely matched that of photosynthesis, in other wavebands the relative spectral responses of some important photomorphogenic reponses were matched. The instrument promises to be useful in the field because it can measure the irradiance in a particular waveband inside a plant canopy in rapidly fluctuating conditions. The development of such "matched" detectors raises some interesting issues. First, what conditions have to be satisfied before a "matched" biological reponse is an acceptable design criterion? Second, when a "matched" instrument is used what precisely is measured?

Woodward matched the relative quantum yield of photosynthesis; it is a well known response and is a good example to examine. The relative quantum yield, Q_{ry}, was defined by McCree (1971,1972) as

$$Q_{ry} = k_1 k_2 (C_L + C_D)/(I \lambda a Q_{my}), \qquad 2.4$$

where k_1 is the constant to convert to micromole/joule, C_L is the CO_2 differential in the light; C_D is the CO_2 differential in the dark, k_2 is the constant to convert to moles/einstein absorbed, Q_{my} is the maximum quantum yield; I is the irradiance; λ is the wavelength and a the absorptance. He determined the relative quantum yield at low irradiance and suggested that the response could be considered to be independent of irradiance. It must be emphasised that before "matching" the detector response the relative quantum yield must be constant at all irradiance levels of practical importance.

Exactly what is measured by a "matched" detector is best determined using a simple example such as a linear relative response (Fig. 2.3). Consider a series of 300 line spectra separated by unit wavelength between 400 and 700 nm, there being 10 photons delivered per second per unit area at each wavelength. The total number delivered is 3000 per second per unit area and it is this that would be measured by a photon detector with a constant photon response. However, with the "matched" detector the photons at the various wavelengths have different degrees of effectiveness. Their effectiveness is scaled relative to the maximum effect; photons

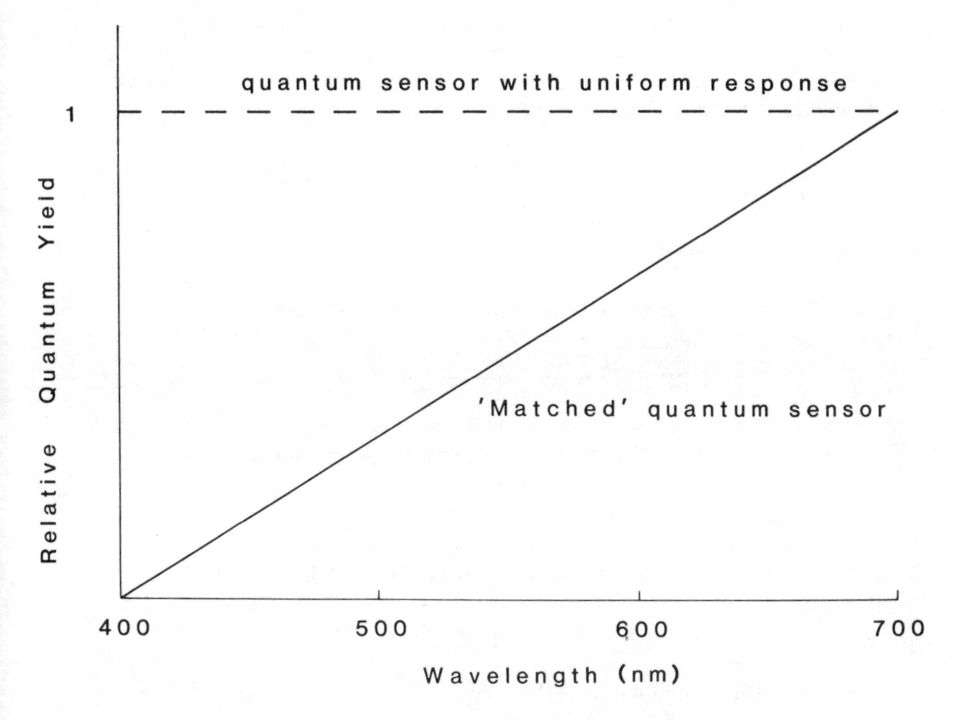

Figure 2.3 A 'matched' detector with a simple linear response.

at 400 nm have no effect and those at 700 nm have full effect. In the example, it can be seen that an average photon has half the effect of a photon at 700 nm, in other words the 3000 photons in the waveband 400-700 nm are equivalent in effect to 1500 photons at 700 nm. Thus the *output* of a "matched" detector may be defined as

$$Output = K\Sigma_{\lambda_1}^{\lambda_2} N_\lambda R(\lambda) \qquad 2.5$$

where $R(\lambda)$ is the relative quantum response (which is dimensionless) at the wavelength λ and N_λ is the number of photons at wavelength λ and K is a detector constant. The output is proportional to a number of photons, equivalent in effect to the same number of photons at the wavelength of the maximum relative quantum yield. The difference between the total number of photons and the "equivalent" number of photons is the number of photons which has no direct effect on the process under consideration. It is interesting to note that the "matched" detector can also be used as an energy sensor, by multiplying the "equivalent" number of photons by the energy of a photon at the wavelength of maximum response.

Accuracy, repeatability and calibration

The theory of instrument and observer errors and their treatment are described by Topping (1955). Woodward and Sheehy (1983) have applied the theory to some problems in environmental biology. It is quite clear that most instrument salesmen and some scientists are unaware of the concepts which underlie statements about the accuracy of an instrument. Hayward (1977) has presented a concise description of the meaning of accuracy in his monograph.

The types of error, often quoted for radiation detectors, and their causes are shown in Table 2.4. The importance of the calibration procedure cannot be overstated; the sub-standard used must have been maintained in good condition and its calibration regularly checked against a standard higher in the calibration chain.

Table 2.4 Types of error quoted for radiation detectors and their meaning.

Type of error	Causes
Systematic error	The chain of uncertainties between the international standardand the sub-standard used in the actual calibration
Cosine error	Departures from the ideal cosine response
Azimuthal error	Changes in the output when the azimuthal angle is varied
Spectral error	Deviation from the ideal spectral response
Temperature error	Changes in output resulting from changes in temperature

The objective of the calibration procedure is to ensure that values recorded using the instrument are as independent of the instrument as is possible i.e. an absolute measure not influenced by the characteristics of the instrument. Thus, it is important to ensure that the effects of environmental conditions likely to be encountered in the measuring situation are accountable. A detailed example of the calibration of a radiometer is given by Wyatt (1978).

Lasers and Optoacoustics

The word laser was originally an acronym for a process called "light amplification by the stimulated emission of radiation", but it has come to mean a device which produces coherent monochromatic radiation. Because the output from a laser is stimulated in a controlled manner, the waves of laser radiation are of uniform length, coordinated in space and time and their intensity per wavelength is very high. Lasers which emit radiation continuously, continuous beam, are usually much less powerful than those which produce radiation in short duration pulses. It is possible to produce laser emission over the waveband 100-30000 nm by tuning different types of laser systems (Gelbwachs, 1977; Kelley, 1977). The development of tuned lasers has resulted in the development of laser based spectroscopic techniques for the detection of atmospheric pollution. An excellent review of this subject has been presented by Patel (1978). There are four basic categories of laser:

(1) solid-state, such as the ruby laser which require optical stimulation;

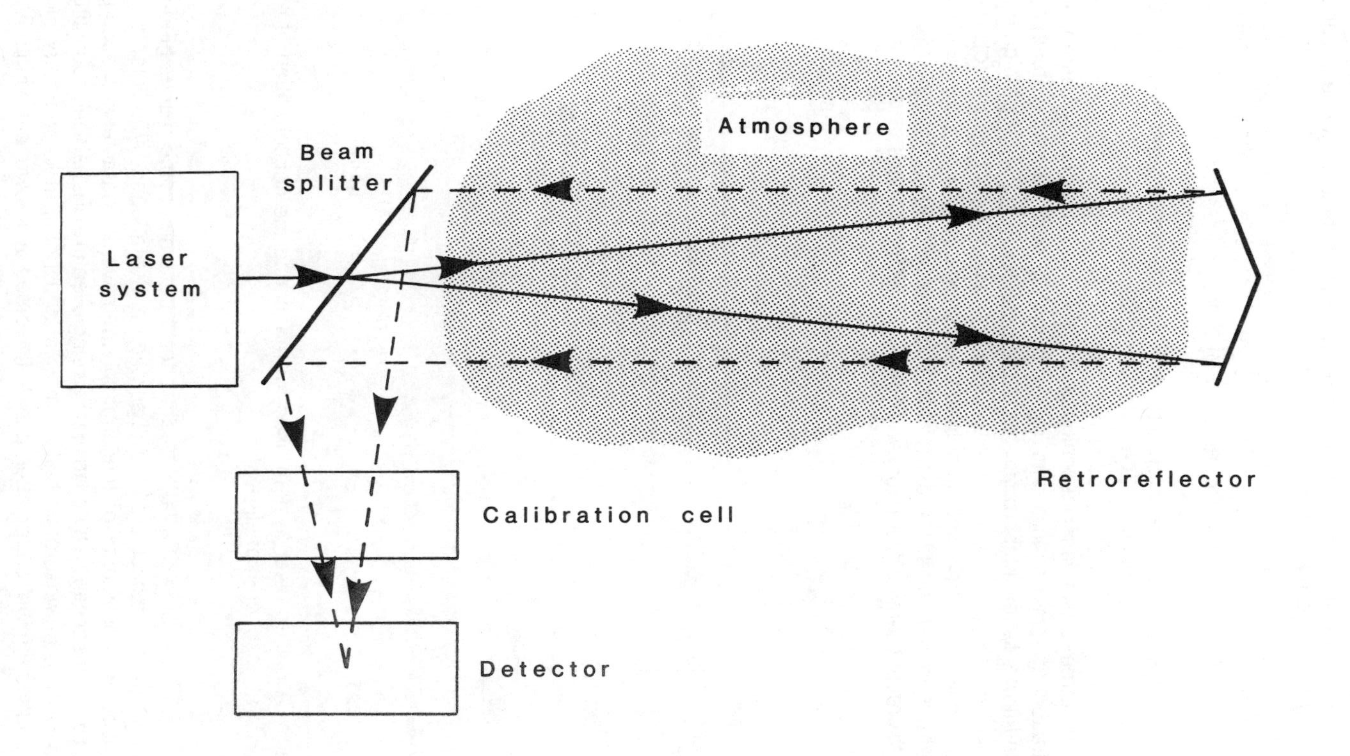

Figure 2.4 A continuous beam laser used for measuring the concentration of minor constituents of the atmosphere.

(2) semiconductor, in which the hole electron pairs recombine to produce quanta; the production of light depends on the current flow through the semiconductor and this can easily be modulated;

(3) gas lasers, such as the carbon dioxide laser, require a voltage gradient and stimulation at low frequency; they are very efficient;

(4) liquid lasers; such as the organic dye types which are stimulated using other lasers; they are easily tuned over the waveband 300-1200nm.

There are three basic laser techniques for the measurement of atmospheric concentrations of minor constituents: (a) a continuous beam laser traversing a long path with mirrors to reflect the beam back to a detector (Fig. 2.4); (b) a high energy pulse laser and the analysis of back scattered radiation (Fig. 2.5); and (c) passing samples of the atmosphere into an optoacoustic chamber. The principles underlying each method are very simple and each has certain advantages (Patel, 1978). I shall attempt to present some of the theory and outline some applications.

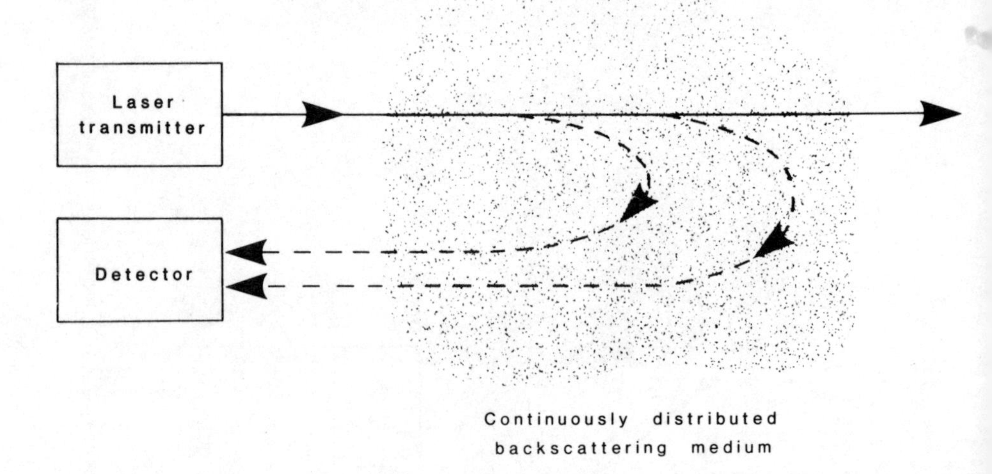

Figure 2.5 A pulsed laser system used for measuring the concentration of minor atmospheric constituents.

The decrease in energy of a laser beam after it traverses some pathlength in the atmosphere gives a measure of the absorption and scattering by pollutants. Most minor gaseous constituents of the atmosphere have a characteristic absorption spectrum and a laser can be tuned to match a single feature of a particular constituent. Tuning a laser to an absorption feature at λ_1 and just off it at λ_2 and detecting the energy change in the beam after it has traversed a known pathlength in the atmosphere provides two measurements (Fig. 2.6). The difference between the two

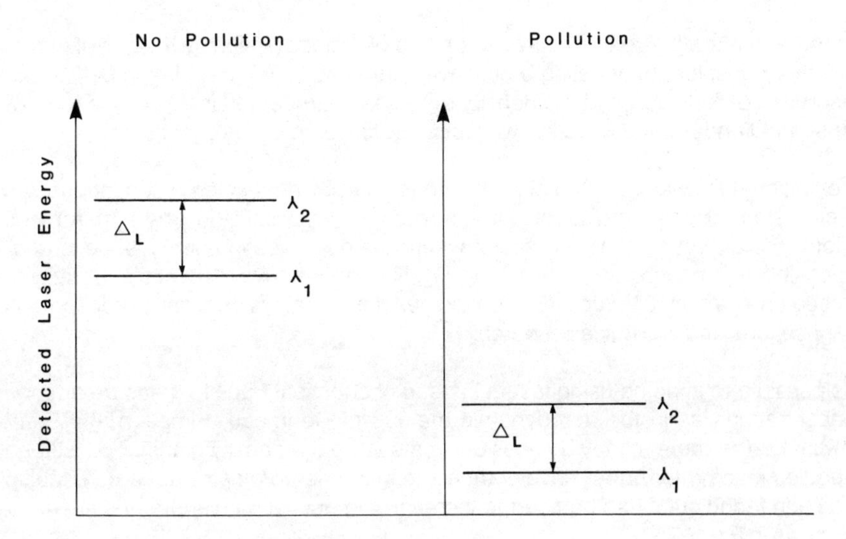

Figure 2.6 The use of a 'differential' laser system to measure gaseous concentration, the wavelengths λ_1 and λ_2 are close, with λ_1 being strongly absorbed by some gaseous component of the atmosphere.

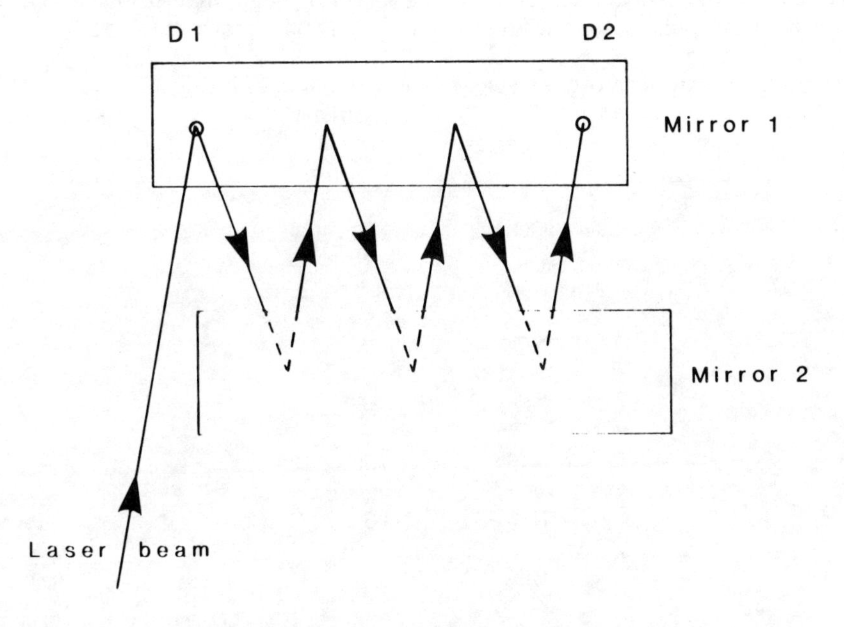

Figure 2.7 Increasing the optical path using multiple reflection.

measurements will give an accurate measure of the absorption by the gas of interest, the other constituents affecting both wavelengths equally. The relation between the absorption of radiation and the density of the gas can be obtained using a tube with transparent ends containing a known concentration of gaseous constituent.

The National Physical Laboratory (NPL) has recently developed a continuous wave laser system; it emits radiation in a very narrow bandwidth towards a mirror which reflects the energy back to a detector where the decrease in energy is measured. It is possible to measure CO_2, H_2O, NH_3, O_3, N_2O and many pollutants; sensitivity can exceed one part in 10^8. Such laser based gas detection systems have great potential in agricultural and ecological research.

It is possible to imagine using lasers instead of chemical traps to measure ammonia concentrations as in the aerodynamic method of Denmead, Simpson and Freney (1977). Furthermore, it may be possible to measure the density of minor constituents in eddies leading to further developments in micrometeorological research. Multiple reflection techniques can be used to increase the optical pathlength and sensitivity, as shown in Fig. 2.7.

The NPL has also developed the technique in which a short duration laser pulse is launched into the atmosphere. The laser is tuned to and just off some desired frequency and the radiation back-scattered by the atmosphere is measured. The time interval between the launch and receipt gives a measure of the pathlength traversed and so the distribution of the constituent along the optical path can be determined. The advantages of laser systems are (a) their high resolutions, which minimises interference and (b) their rapid time responses, which means that fluctuations can be monitored. The present sensitivity of such laser systems is shown in Table 2.5.

Table 2.5 The sensitivity of gas laser systems currently used in Britain.

Species	Detection limit ($ppm\ m^{-1}$)
CO	3
N_2O	1.5
NO_2	25
NO	35
H_2S	750
NH_3	6
CO_2	<1
H_2O	<1

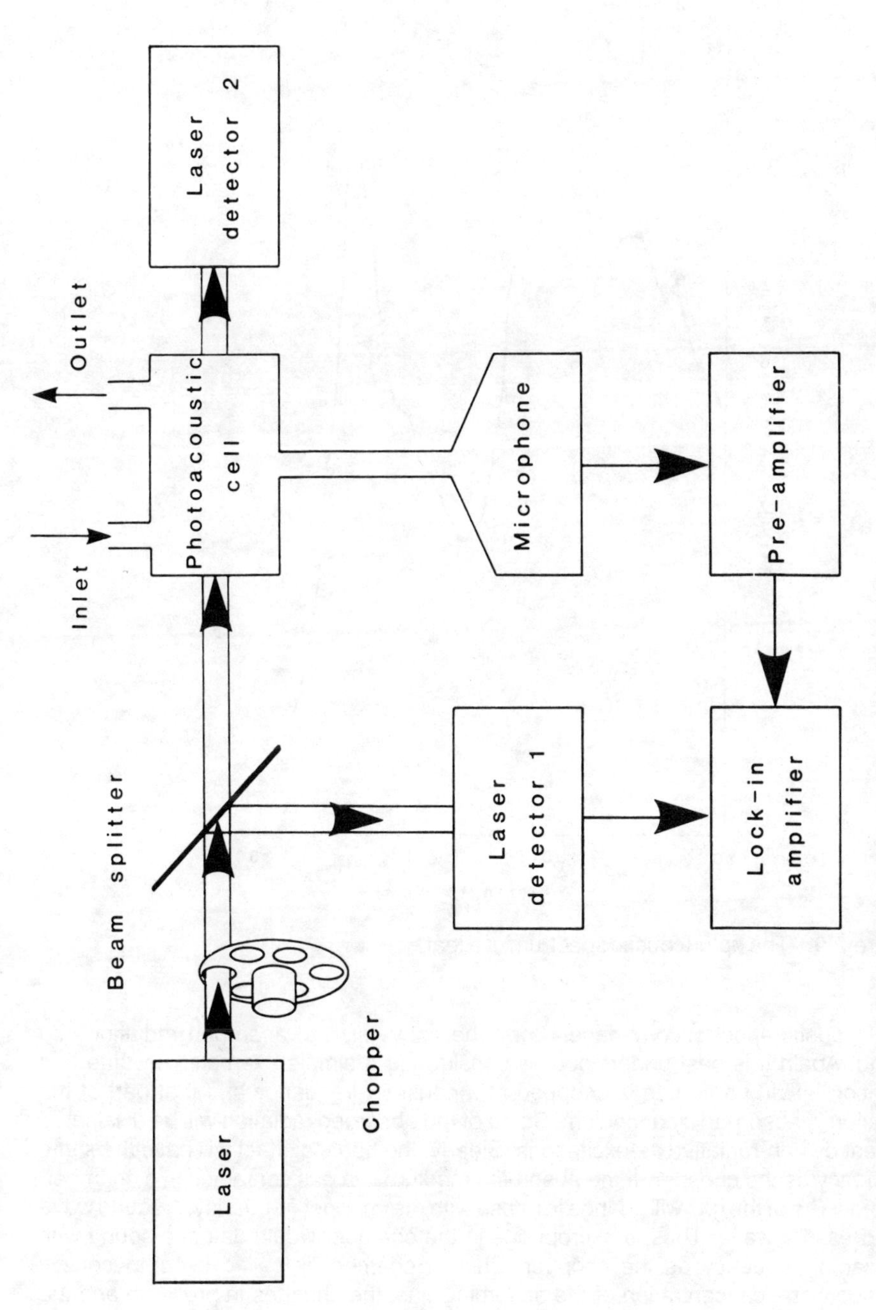

Figure 2.8 A laser system for use in optoacoustic spectroscopy.

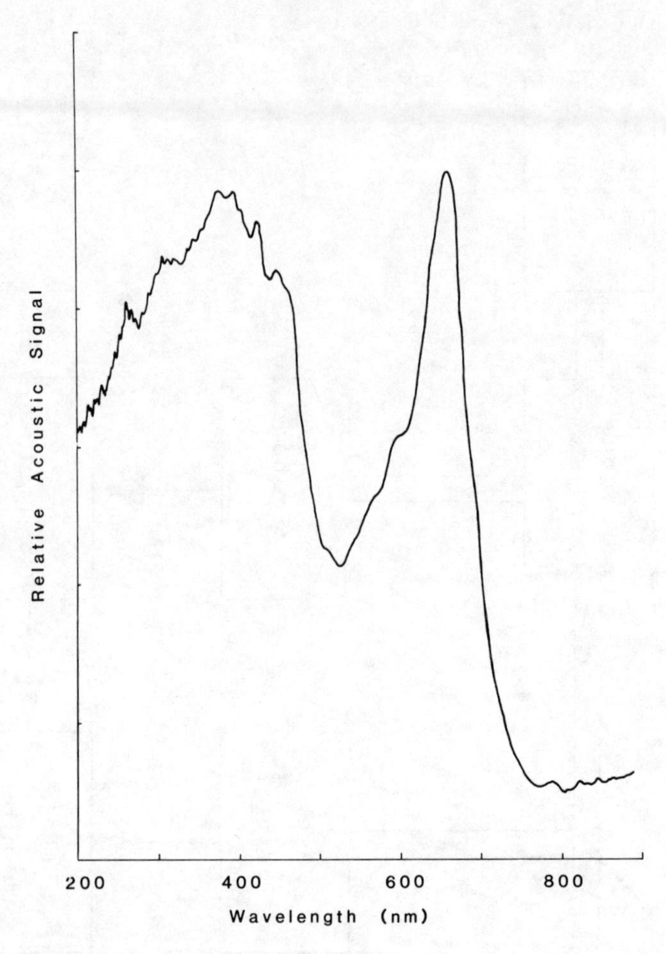

Figure 2.9 The optoacoustic spectrum of a leaf.

Optoacoustic spectroscopy depends on the conversion of absorbed radiation into sound. Again it is best understood by considering a simple example. Assume that radiation leaving a source is chopped at a constant frequency and that part of the radiation is absorbed by a medium. Some of the absorbed radiation will be dissipated as heat by non-radiative de-excitations. Clearly, the heating effect will have the same frequency as the chopper. If the absorbing medium is a gas contained in a chamber, the pressure of the gas will change in phase with the chopper frequency. A sound wave is a pressure wave. Thus, a microphone in the chamber would detect a sound with the same frequency as the chopper. Such a chamber is called an optoacoustic chamber. The concentration of the absorbing gas, the changes in pressure and the sound level are closely related and so the system is easily calibrated. The technique can also be applied to solids and liquids. By varying the wavelength of the incident radiation an optoacoustic spectrum can be assembled (Fig. 2.8).

Optoacoustic techniques can be applied to plant material as shown in the optoacoustic spectrum of a leaf in Fig. 2.9. The technique can be applied to study surface characteristics as well as the progress of chemical reactions. Methodology for the application of the photoacoustic technique to photosynthesis research has been presented by Lasser-Ross, Malkin and Cahen (1980). Some recent developments have been desribed by O'Conner and Diebold (1983). It is interesting to note that the longwave radiation emitted by the absorbing medium will have the same frequency as the chopper. The detection of modulated thermal radiation from leaves may help to assess the efficiency of the absorbing systems in enclosed situations. Finally, it is clear that some research effort to assess the application of laser techniques in agricultural and ecological research is overdue.

References

Anderson, M.C. (1971). Radiation and crop structure. *In* Plant Photosynthetic Production: Manual of Methods, eds. Z. Sestak, J. Catsky, P.G. Jarvis, pp.412-466. The Hague: Dr W. Junk.

Baille, A., Guicherd, R. & Mermier, M. (1978). Study of possibility to use a selective pyranometer in agronomy and atmospheric pollution. Annls. agron., **29**(1), 59-78.

Bell, C.J. & Rose, D.A. (1981). Light measurement and the terminology of flow. Plant, Cell & Environ., **4**, 89-96.

Betts, D.B. (1965). The Spectral Response of Radiation Thermopiles. J. scient. Instrum., **42**, 243-247.

Biggs, W.M., Edison, A.R., Eastin, J.D., Brown, K.W., Maranville, J.W. & Clegg, M.D. (1971). Photosynthesis light sensor and meter. Ecology, **52**(1), 5-131.

Bonhomme, R. & Chartier, P. (1972). The interpretation and automatic measurement of hemispherical photographs to obtain sunlit foliage area and gap frequency. Israel J. agric. Res., **22**(2), 53-61.

Bunnik, W.J.J. (1978). The multispectral reflectance of shortwave radiation by agricultural crops in relation with their morphological and optical properties. Meded. LandbHoogesch. Wageningen, 78-81.

Burgess, M.D. & Cox, L.M. (1975). A bipolar analog integrator for use with net radiometers. Agric. Meteorol., **15**, 385-391.

Campbell, G.S. (1974). A micropower electronic integrator for meteorological applications. Agric. Meteorol., **13**, 399-404.

Campbell, G.S., Mugas, J.N. & King, J.R. (1978). Measurement of long-wave radiant flux in organismial energy budgets: a comparison of three methods. Ecology, **59**(6), 1277-1281.

Chappell, A. (1976). Optoelectronics: Theory and Practice. Texas Instruments Ltd.

Constantine, T.R. (1979). A simplified tube solarimeter. A.A.B. 75th Anniv. Meeting Conversazione. 18th Sept. 1979.

Denmead, O.T., Simpson, J.R. & Freney, J.R. (1977). A direct field measurement of ammonia emission after injection of anhydrous ammonia. J. Soil Sci. Soc. Am., **41**, 1001-1004.

Eckardt, F.E., Methy, M. & Sauvezon, R. (1969). Un spectroradiometre-fluxmetre de photons pour l'etude du climat radiatif au sein de l'ecosysteme. Oecol. Plant., N., 267-297.

Fitter, D.J., Knapp, P.H. & Warren Wilson, J. (1980). Stand structure and light penetration. IV. A sensor for measuring photosynthetically active radiation. J. appl. Ecol., **17**, 183-193.

Fritschen, L.J. & Gay, L.W. (1979). Environmental Instrumentation. New York, Heidelberg, Berlin: Springer-Verlag.

Gelbwachs, J.A. (1977). Tunable radiation sources in the ultra-violet and visible spectral regions (0.1 - 1.0 μm). *In* Optoacoustic spectroscopy and detection, ed. Yoh-Han Pao, N.Y., San Francisco, London: Academic Press.

Grace, J. (1971). The directional distribution of light in natural and controlled environment conditions. J. appl. Ecol., **8**(1), 155-164.

Hayward, T.J. (1977). The state of the art of accuracy assessment: total confusion. Control & Instrum., Oct. 1977, pp.38-41.

Hirota, O., Takeda, T. & Saito, Y. (1975). Studies on the utilisation of solar radiation by crop stands. 1. A photosynthetically active solarimeter designed for use in a leaf canopy. Proc. Crop Sci. Soc. Japan. **44**(3), 357-363.

Hooper, B.E. (1976). The application of hemispherical photography to cereal canopies. J. appl. Ecol., **13**(2), 555-561.

Incoll, L.D., Long, S.P. & Ashmore, M.R. (1977). SI Units in publications in plant science. Commentaries in Plant Science, no. 28. Curr. Adv. Plant Sci., **9**, 331-332.

Kelley, P.L. (1977). Tunable infra-red laser sources for optoacoustic spectroscopy. *In* Optoacoustic spectroscopy and detection, ed. Yoh-Han Pao, N.Y., San Francisco, London: Academic Press.

Klabzuba, J. (1975). Construction of a tube solarimeter suitable for the measurement of some components of solar radiation in the stand of farm crops. Sb. vys. Sk. zemed. Praze, Faculta agronomicka, rada A, 1975, 171-182.

Kubin, S. (1971). Measurement of radiant energy. *In* Plant Photosynthetic Production: Manual of Methods. eds. Z. Sestak, J. Catsky, P.G. Jarvis, pp.702-765. The Hague: Dr W. Junk.

Lasser-Ross, N., Malkin, S. & Cahen, D. (1980). Photoacoustic detection of photosynthetic activities in isolated broken chloroplasts. Biochim. biophys. Acta, **593**, 330-341.

Luxmoore, R.J., Millington, R.J. & Aston, A.R. (1971). Modified tube solarimeter for additive and net measurements of visible, infra-red and solar radiation. Agron. J., **63**, 329-331.

McCree, K.J. (1971). The action spectrum, absorptance and quantum yield of photosynthesis in crop plants. Agric. Meteorol., **9**, 191-216.

McCree, K.J. (1972). Test of current definitions of photosynthetically active radiation against leaf photosynthesis data. Agric. Meteorol., **10**(3), 443-453.

Muchow, R.C. & Kerven, G.L. (1977). A low cost instrument for measurement of photosynthetically active radiation in field canopies. Agric. Meteorol., **18**, 187-195.

O'Conner, M.T. & Diebold, G.J. (1983). Chemical amplification of optoacoustic signals. Nature, Lond., **301**, 321-322.

Oquist, G., Hallgren, J.E. & Brones, L. (1978). An apparatus for measuring photosynthetic quantum yields and quanta absorption spectra of intact plants. Plant, Cell and Environ., **1**, 21-27.

Palmer, J.W. (1980). The measurement of visible irradiance using filtered and unfiltered tube solarimeters. J. appl. Ecol., **17**, 149-150.

Patel, C.K.N. (1978). Laser detection of pollution. Science, N.Y., **202**(4364), 157-173.

Reyenga W. & Dunin, F.X. (1975). An assembly for measuring components of the radiation balance. Aust. CSIRO Div. Plant Ind. Field Stn. Rec., **14**, 17-23.

Sheehy, J.E. & Chapas, L.C. (1976). The measurement and distribution of irradiance in clear and overcast conditions in four temperate forage grass canopies. J. appl. Ecol., **13**, 831-840.

Shibles, R. (1976). Terminology pertaining to photosynthesis. Committee Report. Crop Sci., **16**, 437-439.

Sinclair, T.R. & Lemon, E.R. (1974). Penetration of photosynthetically active radiation in corn canopies. Agron. J., **66**, 201-205.

Smith, R.A., Jones, F.E. & Chasmar, R.P. (1968). The detection and measurement of infra-red radiation. Oxford: Clarendon Press.

Steven, M.D. (1977). Standard distribution of clear sky radiance. Q. Jl. R. met. Soc., **103**, 457-465.

Steven, M.D. & Unsworth, M.H. (1980). The angular distribution interception of diffuse solar radiation below overcast skies. Q. Jl. R. met. Soc., **106**, 57-61.

Szeicz, G. (1974). Solar Radiation for Plant Growth. J. appl. Ecol., **11**(2), 617-636.

Topping, J. (1955). Errors of observation and their treatment. London, Reinhold, New York: Chapman and Hall.

Williams, B.A. & Austin, R.B. (1977). Short note: An instrument for measuring the transmission of shortwave radiation by crop canopies. J. appl. Ecol., **14**, 987-991.

Woodward, F.I. (1983). Instruments for the measurement of photosynthetically active radiation and red, far-red and blue light. J. appl. Ecol., **20**, 103-115.

Woodward, F.I. & Sheehy, J.E. (1983). Principles and measurements in environmental biology. London, Boston, Durban, Singapore, Sydney, Toronto, Wellington: Butterworths.

Woodward, F.I. & Yaqub, M. (1979). Integrator and sensors for measuring photosynthetically active radiation and temperature in the field. J. appl. Ecol., **16**, 545-552.

Wyatt, C.L. (1978). Radiometric calibration. Theory and methods. N.Y., San Francisco, London: Academic Press.

Additional useful refererences

Annotated Bibliographies, Commonwealth Bureau of Pastures and Field Crops. (1966-76) no. G443 (2), 16 pp. and (1977-80) no. G443 (2) A, 34 pp.

Sagar, J.C., Edwards, J.L. & Klein, W.H. (1982). Light energy utilisation efficiency for photosynthesis. Transactions of the ASAE, **25**, 1737-1746.

Chapter 3: The measurement of carbon dioxide in air

P. G. Jarvis and A. P. Sandford
Department of Forestry and Natural Resources,
University of Edinburgh, Darwin Building, King's Buildings,
Mayfield Road, Edinburgh EH9 3JU, Scotland.

Introduction

Carbon dioxide is measured in air for the purposes of describing the changes in atmospheric content (e.g. Keeling *et al.*, 1976) and the exchange fluxes between the atmosphere and soil, plants, animals and vegetation (e.g. Sestak et al., 1971). A number of chemical and physical methods have been used for measuring CO_2 in air in the past (see Sestak *et al.*, 1971) but today infra-red gas analysis is by far the most widely used method and through the adoption of new technology is likely to retain this position. Consequently, this chapter is concerned exclusively with infra-red techniques.

Initial development of infra-red gas analysers was based on the detector invented by Luft in the 1940s and the first generation of commercial instruments began to be used in environmental and biological applications in the late 1950s. Sestak *et al.* (1971) provided a comprehensive review of the state of the art in 1970. In this chapter we concentrate on developments since then. These fall broadly into two categories: improved engineering of the concepts in use in 1970, in conjunction with the introduction of new technology, such as solid state choppers and detectors; and practical and theoretical developments regarding instrument calibration and the calculation of fluxes. In addition simultaneous developments of accessories, such as mass flow meters, have led to evolution in the design of gas exchange systems.

Infra-red gas analysers

Heteroatomic gases such as CO_2 absorb radiation in particular wavebands (Table 3.1). Infra-red gas analysers measure the density of a particular gas present in a carrier by the reduction in transmission of infra-red radiation at one or more of these wavebands.

Infra-red gas analysers can be broadly classified into *dispersive* and *non-dispersive* types. In dispersive infra-red (DIR) analysers the transmission of *monochromatic* radiation by the gas mixture is measured. Dispersive infra-red analysers consist essentially of a through-flow gas cell placed in the optical path of an infra-red spectrometer. By scanning a spectrum of wavelengths from 2.5 to 14.5 μm, the composition of a mixture of gases can be determined, and the amount of a particular species present found from the reduction in transmission at one or more wavelengths.

Table 3.1 Main absorption bands of carbon dioxide in the intermediate infra-red region (from Janac *et al.*, 1971).

Peak wavelength λ (μm)	Wavenumber ν (mm^{-1})	Band Intensity S (Pa m^{-2})	Average absorption coefficient k (MPa m^{-1})
2.69	371.6	5.3±1.0	0.53±0.10
2.77	360.9	3.6±0.8	0.31±0.07
4.26	234.9	246.7±39	22.9±3.66
14.99	66.7	32.6±8.9	3.1±0.84

This method is particularly suitable for complex mixtures of gases and has been developed for the measurement of trace amounts of a gas by the use of very long optical paths of up to 132 m!

In non-dispersive infra-red (NDIR) analysers the transmission of *broad-band*, unselected infra-red radiation is measured. The measurement is made selective to a particular species of gas by filling the detector with the gas in question and/or by the use of filters in the optical path. Gases other than the one being measured may absorb radiation in a waveband that overlaps one of the absorbing wavebands of the measured gas. This leads to cross-sensitivity of the detector to the interfering gas. For example, absorption by water vapour overlaps that by CO_2 at 2.7 μm. Filters are used to minimise the consequence of this interference. They may be either closed, gas-filled cuvettes or narrow-band, optical thin-film filters.

NDIR analysers are much more commonly used for the routine analysis of a single species of gas such as CO_2 than DIR analysers, because they are generally cheaper and more robust since they do not require the high precision optics of an infra-red spectrometer. However, there are advantages in measuring absorption by a mixture of gases over a spectrum of wavelengths, notably in accounting for cross-sensitivity to other species.

Principle of infra-red gas analysers

The electrical output from the detector depends on the flux of radiation that is transmitted by the gas-containing cell and is absorbed in the detector. In principle transmission of radiation of a particular wavelength through a cell follows the Beer-Lambert law (Janac, 1970) and is reduced in relation to the density of CO_2 in the cell (ρ) and the length of the cell (l). Thus the transmission of a cell for radiation at wavelength (λ) is:

$$\tau(\lambda) = Q_t/Q_o = \exp(-\rho l k(\lambda)) \qquad 3.1$$

where k(λ) is the extinction coefficient for radiation of wavelength λ from the source (Q_o) and Q_t is the flux of radiation that is transmitted by the cell. The absorptance of radiation of wavelength λ in the cell is:

$$\alpha(\lambda) = 1 - \tau(\lambda) \qquad 3.2$$

Absorption of radiation in a waveband by a heteroatomic gas occurs at distinct, very narrowly defined wavelengths or absorption lines that correspond to the rotational states of the molecules. Thus to obtain the absorptance by CO_2 in the broad 4.1 to 4.3 μm absorption band (α), equation 3.2 is integrated over the full number (c. 61) of absorption lines.

In a typical dual-beam analyser, the output from the detector (V) is proportional to the difference between the fluxes of radiation transmitted by the reference and sample cells and, for a constant source strength, is therefore proportional to the difference in transmittance of the reference cell (τ_r) and the sample cell (τ_s). In an absolute analyser the transmittance of the reference cell is maximal and fixed and consequently detector output is proportional to the absorptance of the sample cell (α_s), i.e. $V \simeq (\tau_r - \tau_s) = 1 - (1 - \alpha_s) = \alpha_s$.

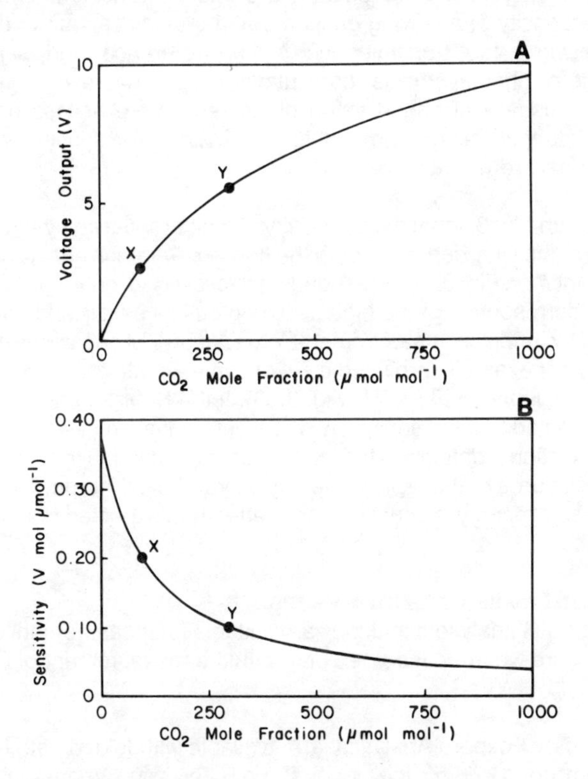

Figure 3.1 Diagrams showing the output (a) and the sensitivity (b) of a typical, non-linearised, differential analyser as a function of mole fraction.

Fig. 3.1a shows that the relation between τ_s and ρl is curvilinear as would be expected from equations 3.1 and 3.2 although in practice it is not exponential, as predicted, possibly because of the distribution of the absorption lines and the effect of gas density or other variables on their width. Multiple reflections of the beams within the reference and sample cells of analysers that do not have imaging optics may also contribute to departure from the exponential.

It is clear from Fig. 3.1b that the sensitivity of an absolute analyser [$S = dV/d(\rho l)$] is a function of the CO_2 density in the sample cell (ρ_s), for a particular cell length. If either ρ_s or l is substantially increased, the response becomes increasingly non-linear and the sensitivity falls. Conversely if l is very short, the signal to noise ratio may become too small. Hence, l is chosen for a particular measurement range of ρ_s to be as short as possible consistent with measurement accuracy. In practice with contemporary analysers, cell lengths of about 250 mm are used in the μmol mol^{-1} range and of about 25 mm in the 1-10% range.

In a differential analyser, the transmittance of the reference cell is also reduced in relation to the density of a flowing gas, so that the electrical output depends on the difference between the transmittance of the reference and sample cells. In measurements of photosynthesis, particularly in open gas-exchange systems, the CO_2 density is usually arranged to be higher and more or less unvarying in the reference cell, so that V is proportional to ($\alpha_r - \alpha_s$) and the sensitivity [$S = dV/d(\rho_r - \rho_s)$] is proportional to $d(\alpha_r - \alpha_s)/d(\rho_r - \rho_s)$.

It is apparent from Fig. 3.1 that the sensitivity of a differential analyser, with cell length l, is a function of the CO_2 density of both the flowing reference gas (ρ_r) and the sample gas (ρ_s). At point X on Fig. 3.1, corresponding to a low ρ_r, $dV/d(\rho_r - \rho_s)$ is twice as large as at point Y, corresponding to a high ρ_r. The practical significance of this has been demonstrated by empirical measurements on a number of occasions. Bierhuizen and Slatyer (1964), for example, provided a set of curves relating V to ($\rho_r - \rho_s$) at different values of ρ_r (see Janac *et al.*, 1971, Fig. 3.33). If the Beer-Lambert law held exactly, it would be expected that the sensitivity of the differential analyser would be a double exponential function containing the reference and sample gas densities, but in practice this relation is better represented by a hyperbola (Thorpe, 1978; Bloom *et al.*, 1980). As a result linearity is usually rather better than expected from the exponential equation.

Dispersive infra-red gas analysers

It is a feature of DIR analysers that they are suitable for measurement of a wide range of gasses but here we are concerned only with the measurement of CO_2 at the 4.26 μm waveband.

Gas cells made by Foxboro Analytical are available with folded path lengths ranging from 100 mm to 132 m, the longest cells allowing measurement of trace concentrations to CO_2 of 0.05 μmol mol^{-1}. These long folded path lengths are achieved by multiple reflections between gold-plated mirrors. The volume of these gas cells is considerable (2500-95000 cm^3) and this imposes a long response time in continuous flow applications. The electronic response time is of the order of 100 ms, and in principle the walls of the gas cell could be partially cut away for open path measurement. All the cells can be attached to a wide range of single and dual beam, laboratory infra-red spectrometers (e.g. Beckman, Perkin-Elmer, Bausch and Lomb,

Carl Zeiss, Nicolet and Pye Unicam) as well as to the range of portable and laboratory single beam, Miran infra-red spectrometers made by Foxboro Analytical (see Table 3.3). The more elaborate Miran spectrometers contain a circular variable filter that gives monochromatic radiation between 2.5 and 14.5 μm (±0.05 to ±0.25 μm) whereas in the simpler portable spectrometers, narrow-band filters are used and measurement is restricted to one or two wavebands (Table 3.3). The distinction between DIR and NDIR analysers breaks down when the infra-red spectrometer is replaced by fixed monochromatic radiation in one particular waveband. Such an analyser is then indistinguishable in principle from a single beam NDIR analyser with a broad-band infra-red source and a highly selective thin-film filter in the optical path.

Non-dispersive infra-red gas analysers

Almost all NDIR analysers possess the following components, though not necessarily in the order given: infra-red source(s), chopper, reference and sample cells, filters and infra-red detector(s). A comprehensive review with illustrations of a number of configurations then current was given by Janac *et al.* (1971) and it seems unnecessary to go over the ground in such detail again. A configuration typical of several contemporary analysers is shown in Fig. 3.2. A number of variations on this basic pattern are apparent in the analysers currently available commercially, and the last twelve years has seen substantial innovation in different instruments in relation to source, chopper and detector, in particular.

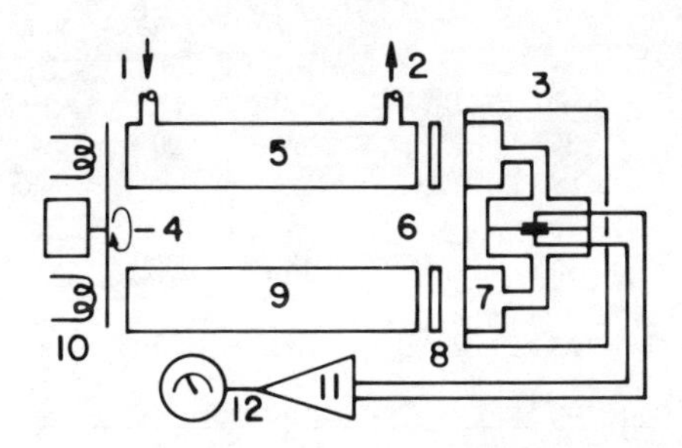

1. Gas in
2. Gas out
3. Luft detector
4. Rotating shutter
5. Analysis cell
6. Metal diaphragm
7. Absorption chamber
8. Optical filters optional
9. Reference cell
10. IR sources
11. Amp
12. Meter

Figure 3.2 A diagram of a conventional dual beam analyser with twin sources and a Luft detector with the absorption chambers in parallel, the Seiger 120.

Analysers readily available today for use in laboratory and field are listed in Tables 3.2 and 3.3, respectively. The distinction is somewhat arbitrary and placement in Table 3.3 is based almost entirely on provision by the manufacturer for direct operation from a battery supply. It will become apparent that one or two of the analysers in Table 3.2

Table 3.2 A list of infra-red gas analysers for laboratory use.

	1	2	3	4	5	6	7	8
min range (μmol mol^{-1})	25	10000	50	100	500	10	–	30
max range (%)	2.5	20	0.05	100	100	100	10	100
no. of ranges/inst.	4	1	2	5	3	3	1	4
accuracy (±% fsd)	1	2	1	0.5	1	1	1	1
zero stability (±%/24h)	1	1	1	1	1	1	1	1
90% min response time (s)	3	3	5	0.3	0.5	0.5	0.13	?
warm up time (min)	30	60	30	120	60	60	240	15
Differential, Absolute	D+A	A	D,A	A	A	D,A	A	D,A
detector type	L(p)	p.e.	PbSe	p.e.	L(p)	L(p)	L(p)	L(s)
chopper speed (Hz)	6.7	10	150	80	10	10	90	10
no. of cells	2	1	2	1	2	2	2	2
no. of sources	1	1	1	1	2	2	2	1
filters	i+g	i	i	2i	i,g	i,g	–	–
linearisation	+	±	+	+	+	+	+	+
internal calibration	t.l.	–	+	–	+	+	–	–
thermostat	+	–	–	±	+	+	+	–t.c.
pressure compensation	–	–	–	±	–	–	+	–
case/optics purge	+	+	–	+	+	+	+	?
display	a,d	a,d	d	d	a	a	d	a
output	mV	V	mV	V,mA	V,mA	V,mA	V	V,mA
includes pumps	+	+	+	–	–	–	?	?
temp. tol (°C)	0-40	0-40	0-50	5-50	–1-49	–1-49	15-40	0-40
humidity tol (%RH)	85	85	?	100	90	90	95	?
vibration	intol	tol	tol	tol	intol	intol	intol	?
mains tol (V) (±%)	10	15	?	15	13	13	15	?
mains tol (Hz) (±%)	5	2	?	15	1	1	5	15
std. rack mounting	±	±	±	–	+	+	±	+
weight (kg)	27	10/11	23/46	34	23	28	17	15

1, ADC 225(3); 2, ADC SS; 3, Anarad APPA Series; 4, Anatek 402; 5, Beckman 864; 6, Beckman 865; 7, Beckman LB2; 8, Cosma 6000

Abbreviations: a, analogue; d, digital; fsd, full scale deflection; g, gas filters; i, interference filters; intol, intolerant; L(p), Luft(parallel); L(s), Luft(series); m.flow, mass flow; n.a., not applicable; n.c. nickel cadmium; p.e., pyroelectric; ss, solid state; s.l.a., sealed lead acid; t.c., temperature compensation; t.l., tube length; therm., thermal; tol, tolerant.

possess particular properties that make them suitable for field use despite their requirement for mains voltage. Conversely, several analysers that are particularly satisfactory as field instruments are extrememly useful in the laboratory both in fixed installations and in portable rigs.

Whilst most analysers are sold ready to use, the optical bench and electronics of the

9	10	11	12	13	14	15	16	17	18
30	200	20	20	40	100	200	200	200	30
100	100	0.1	100	100	100	100	50	50	100
3	2	2	3	4	2	2	4	2	4
0.5	1	1	1	0.5	0.5	1	0.5	0.5	1
1	2	2	2	1	2	1	1	1	1
0.3	0.5	0.5	0.5	0.8	5	0.5	1.3	1.5	3.5
?	240	240	240	30	240	15	30	30	60
A	D,A	D,A	A	D,A	A	A	D,A	D,A	D,A
L(p)	L(p)	L(p)	L(p)	L(s)	L(p)	PbSe	L(s)	L(s)	L(p)
10	8.3	8.3	8.3	8.3,125	8.3	50	6.2	6.2	6.6
2	2	2	2	2	2	2	2	2	2
2	2	2	2	1	1	1	1	1	2
–	i	i	i	i,g	i,g	i	i	i	i
–	+	+	±	+	+	+	+	+	+
–	–	–	–	–	–	+	–	–	–
±	+	+	+	+	+	±	+	+	–
–	–	–	–	±	–	–	±	±	–
–	+	+	+	+	+	+	+	+	+
a	a	a	a	a	a	a,d	a,d	a,d	a
V,mA	V,mA	V,mA	V,mA	V,mA	mA	V,mA	mA	mA	V,mA
?	–	–	–	–	+	?	±	+	+
?	0-40	0-40	0-40	5-45	5-45	0-50	0-40	0-45	0-32
?	85	85	85	75	75	95	90	75	99
intol	intol	intol	intol	tol	intol	tol	intol	intol	intol
?	10	10	10	12	12	10	15	10	15
15	0.5	0.5	0.5	4	4	?	2	0.5	20
–	–	–	–	±	±	±	±	±	+
?	15	20	13.5	25/27	12/22	15	18-25	12-20	15-24

9, Cosma 3000; 10, Horiba VIA 300; 11, Horiba VIA 500; 12, Horiba PIR 2000; 13 H & B URAS 3E/G; 14, H & B URAS 7N; 15, IRI 705; 16, Maihak UNOR 4N; 17, Maihak UNOR 6N; 18, Sieger 120

Liston Edwards, Edinburgh Instruments and IRI series 700 are available in kit form for building into one's own apparatus.

Detector developments. In 1971 virtually all infra-red gas analysers made use of the Luft detector and it is clear from Tables 3.2 and 3.3 that a large number do so today. The Luft detector is essentially a differential gas thermometer with two radiation absorption chambers containing the same species of gas that is being measured (i.e. it is made selective by positive filtering). The two chambers are separated by a thin membrane, usually of aluminium alloy or gold 5 to 10 μm thick, that forms one electrode of a diaphragm capacitor. Because the radiation paths are chopped, temperature and pressure fluctuations occur in the detector absorption chambers causing pulsation of the diaphragm with consequent changes in capacitance.

Table 3.3 A list of battery-operated infra-red gas analysers.

	1	2	3	4	5	6
min range (μmol mol^{-1})	30	50	50	1000	1000	2000
max range (%)	100	0.05	0.05	60	100	0.5
no. of ranges/inst.	4	2	2	1	n.a.	2
accuracy (±% fsd)	1	1	1	1	1	3
zero stability (±%/24h)	1	1	1	2.5	0.2	n.a.
90% min response time (s)	2	5	5	3	0.5	5
warm up time (min)	5	3	30	0.2	?	2
Differential,Absolute	A	A+D	D,A	A	A	A
detector type	p.e.	p.e.	PbSe	p.e.	therm.	L(s)
chopper speed (Hz)	18	ss	150	80	ss	8.3
no. of cells	1	1	2	1	1	1
no. of sources	1	1	1	1	2	1
filters	i+g	i	i	i	i	–
linearisation	+	+	+	–	+	–
internal calibration	–	–	+	+	–	–
thermostat	±	–	–	–	–	–
pressure compensation	–	–	–	–	±	–
case/optics purge	+	+	–	–	n.a.	–
display	a,d	d	d	d	n.a.	a
output	V,mA	mV	mV	mV,mA	V,mA	mV
includes pumps	+	+	+	±	n.a.	+
temp. tol (°C)	0-40	0-40	0-50	0-50	0-80	0-40
humidity tol (%RH)	85	85	?	90	99	90
vibration	tol	tol	tol	tol	tol	intol
incl. battery	–	s.l.a.	?	s.l.a.	–	s.l.a.
battery d.c. (V)	10.5-14	12	?	16	15	6
mains tol. (V) (±%)	15	–	?	10	–	10
mains tol (Hz) (±%)	5	–	?	–4,+30	–	5
weight (kg)	9-21	2.8	23	5.5	n.a.	6

1, ADC RF; 2, ADC LCA; 3, Anarad APPA-4; 4, Anatek 404; 5, Edinburgh Instruments; 6, Horiba APBA 210; 7, Leybold Binos II; 8, Liston Edwards; 9, Miran

The gas absorption chambers of the Luft detector can be arranged in parallel or in series. In parallel arrangement, radiation transmitted through the reference gas measuring cell passes into one absorption chamber and radiation transmitted through the sample gas measuring cell into the other. In the series arrangement, radiation transmitted alternately through the reference and sample cells passes into both absorption cells in series (Fig. 3.3). The gas in the first or front absorption chamber primarily absorbs radiation in the centre of the waveband whereas the gas in the rear absorption chamber largely absorbs the tails of the absorption band. The rear chamber is deeper so that in the null condition the pressure pulses in the two chambers

7	8	9	10
50	1000	10	1
100	10	0.5	5
0.5	1	1	2
2	1	1	1
2	2	n.a.	3
0.1	4	0.1	?
2	20	15	0.5
D,A	A	A	A
m.flow	L(s)	p.e.	L(s)
160	ss 5	47	ss 3.3
2	2	1	2
1	2	1	1
±	g	i	i
±	±	–	–
auto	el'nic	–	–
±(t.c.)	–	–	+
±	–	–	–
–	n.a.	–	–
a,d	n.a.	a	a
V,mA	V	V	mA
+	n.a.	+	–
0-45	0-50	?	0-45
95	?	?	90
tol	tol	tol	intol
–	n.a.	n.c.	n.c.
11.5-35	±15	7.2	12
20	–	–	–
5	–	–	–
6.5	n.a.	8.2	7.4

101; 10, Maihak SIFOR 2. Abbreviations: see Table 3.2

balance. In comparison with the parallel arrangement, this leads, in general, to less cross-sensitivity to other gases, a more stable zero setting and smaller changes in sensitivity as a result of variations in barometric pressure. In 1971 the only analyser made on a commercial scale with the radiation absorption chambers in series was the UNOR 2 (Maikak A.G.). Today, at least as many commercial instruments make use of the series arrangement as the parallel one (Tables 3.2 and 3.3).

One of the major problems of the Luft detector is 'microphony', that is noise that arises from spurious vibration of the diaphragm. Most Luft detectors incorporate a small bore

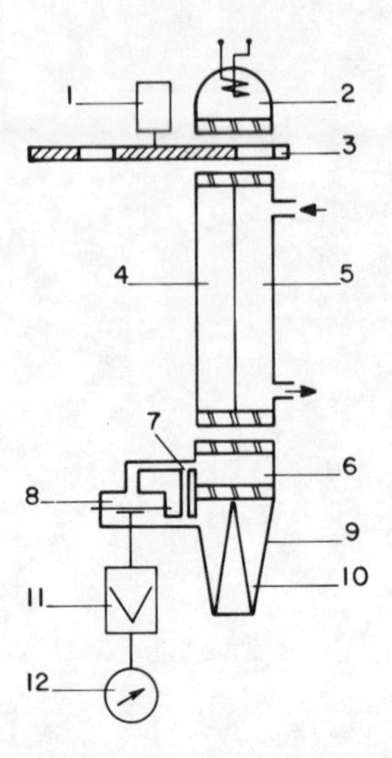

Figure 3.3 A diagram of a single source dual beam analyser with a Luft detector with the absorption chambers in series, the UNOR 4N.

capillary between the two absorption chambers. The bore is chosen to allow pressure equilibration between the chambers at frequencies below the chopper frequency, thus reducing sensitivity to low frequency vibration (see Janac *et al.*, 1971). Narrow bandwidth electronic filters (e.g. as used by Liston Edwards, URAS 3) and phase-selective rectification (e.g. URAS 3, Binos II) also reduce sensitivity to vibration but cannot eliminate it completely. It is apparent from Tables 3.2 and 3.3 that several manufacturers have adopted alternative detectors that try to avoid this problem. The Binos II (Fig. 3.4) utilises a mass flow sensor between an absorption chamber and a compensation chamber and analysers made by ADC (RF, SS and LCA), Anarad and IRI have solid state detectors. Such detectors reduce the sensitivity of the instrument to vibration and consequently are of particular interest for field systems. The IR-700 series makes use of the small size of solid state detectors to analyse two component gases simultaneously in the same sample (e.g. CO_2 and H_2O) by placing two independent filters and detectors in the focus of a mirror that collects the radiation transmitted through the reference and sample cells.

Filter developments. Gas filters *absorb* radiation in *all* the absorption bands of the

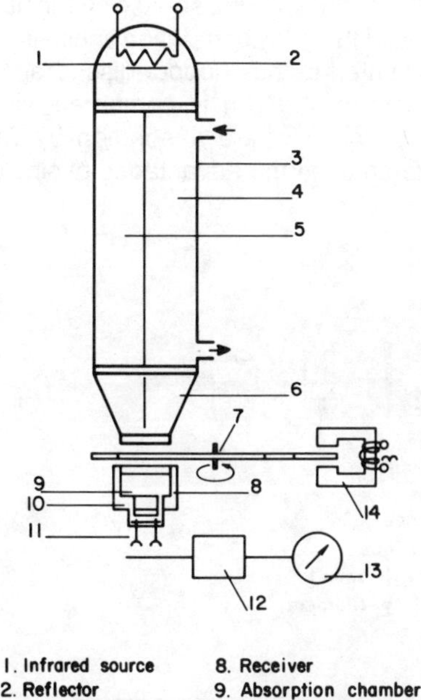

Figure 3.4 A diagram of a single source dual beam analyer, the Binos II. The detector has a single absorption chamber that communicates with a compensation chamber via a mass-flow sensor.

gas contained. Much of the radiation in the absorption bands of an interfering gas can, for example, be filtered out before reaching the detector by a gas filter filled with the interfering gas. Thin-film filters usually *transmit* in *one* particular waveband at which absorption by other, interfering gases is minimal. For example, selection of the 4.26 μm waveband for CO_2 (Table 3.1) greatly reduces cross-sensitivity to water vapour by eliminating transmission of the 2.7 to 2.8 μm waveband in which both water vapour and CO_2 have absorption peaks. Consequently, thin-film filters have become widely used since 1971. However, neither gas filters nor thin-film filters completely eliminate cross-sensitivity to water vapour and it is not uncommon, therefore, to find them both used in combination (e.g. ADC 225, URAS 3, Beckman 865).

In some single beam instruments, the filters are mounted on the chopper. The ADC RF has two gas-filled cuvettes set into the chopper disc, one containing CO_2 the other a neutral gas (Fig. 3.5b). There is also a thin-film filter in front of the sample cell. The CO_2-containing cuvette filters out some of the radiation in the CO_2-absorbing waveband so that when the filter is in the pathway the detector largely responds to changes in the sample cell other than CO_2 variations. When the neutral filter is present, the detector sees all the changes in the sample cell, including CO_2 variations. In the Anatek analysers and two of the eddy correlation specials described later in this chapter, thin-film filters are mounted in the chopper disc. Transmission of radiation in the CO_2-absorbing waveband at 4.3 μm is compared with transmission in a reference waveband (3.7 to 3.9 μm) where absorption by CO_2 is slight. These developments with filters have enabled the advantages of single beam instruments to be exploited (see below).

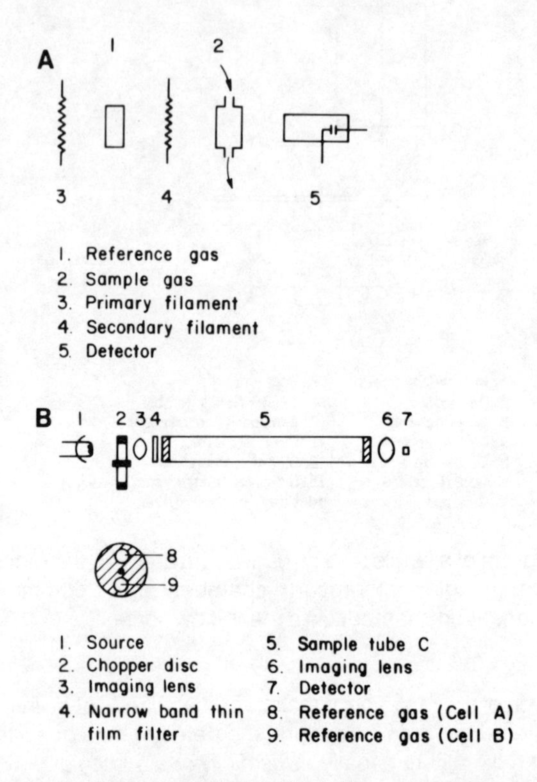

Figure 3.5 Diagrams of two closed path single beam analysers, the Liston Edwards (a) and the ADC RF (b). In both, CO_2 is the reference in the reference cell. In the ADC RF the reference cell is mounted in the the chopper disc and alternates in the optical path with a non-absorbing gas. In the Liston Edwards the filaments are alternately heated.

Source developments. The source of infra-red radiation is generally a coil of nichrome or tungsten wire heated by passage of a current. In some instruments the coil is coated with an oxide layer (e.g. Binos II) or encapsulated in an inert sheath to reduce sublimation of metal from the coil and subsequent contamination of the windows and reflective surfaces surrounding the coil. This acts to reduce calibration drift.

In 1971 most analysers, with the exception of UNOR 2, had two coils, one for the reference and one for the sample gas cells, connected in series in the same circuit. A disadvantage of this arrangement is that differential ageing of one of the coils or its reflector leads to drift in the zero and sensitivity of the analyser. Several instruments now avoid this problem by the use of a single source that either enters the two cells directly (e.g. Binos II, UNOR 4N, 6N, URAS 3, APPA 1-4) or is reflected into the reference and sample cells after being chopped (COSMA Diamant 6000, IRI series 700). The Diamant 6000 design is unusual in that the reflector is on the chopper blade.

Chopper developments. The chopper is usually a mechanically rotated vane or disc that may be found at either the source or the detector end of the optical bench with the drive either in the axis of the bench or placed to one side. In 1971 the chopper in virtually all analysers rotated at between 2 and 10 Hz and was synchronised by the mains frequency. Consequently all analysers were dependent on a highly stable mains frequency and could not be driven from a local generator (Janac *et al.*, 1971). Today a number of instruments have their own internal oscillator to maintain synchronous chopping independent of mains frequency so that the analyser as a whole can tolerate frequency flucuations of up to $\pm 5\%$ (ADC 225, RF, Binos II), $\pm 15\%$ (COSMA), or $\pm 20\%$ (Seiger 120). In some instruments the chopper drive has changed from gears or stepper motors to non-contact methods such as eddy current magnets (Binos II, Horiba VIA 300, VIA 500, PIR 2000). The chopping speed is also much higher in some instruments today than in 1971 (e.g. 160 Hz in Binos II, 50 Hz in IRI series 700). Higher chopping speeds reduce sensitivity to vibration of the detector and may improve the match of detector to amplifier with resulting increase in sensitivity, noise rejection and speed of response. All mechanical choppers result in sensitivity of the analyser to vibration if the frequency of the disturbance is harmonic with the chopper frequency. Thus even instruments with solid state detectors require careful handling and mounting if they have a mechanical chopper. The Liston Edwards 'State of the Art' Gas Sensor (Fig. 3.5a) was the first commercial analyser in which a chopping cycle was generated without the use of a mechanical rotating chopper. There are two infra-red source filaments which are energised alternately and each reach 1200 °C. The new ADC LCA effects chopping by alternately switching the sample gas and CO_2-free air through the single, small volume analysis cell.

The chopper can also be designed to provide an internal calibration standard. In the Binos II, two cutouts in the periphery of the chopper wheel produce a test impulse, the size of which depends on the infra-red source strength and the sensitivity of the detector. Amplification of the signal is automatically adjusted twice every rotation to maintain the test impulses at a fixed height. Consequently amplification of the signal is increased if the sample gas density increases, improving linearity. Long term stability is also improved because changes in the infra-red source and detector are automatically compensated for (Shunck, 1978).

Gas pathway developments. The majority of analysers contain parallel reference and sample cells between the source(s) and detector, exceptions being certain single beam path instruments (ADC RF, SS and LCA, Liston Edwards, Anatek 402, 404) and one or two others with unusual configurations. The COSMA Rubis 3000, for example, has a central Luft detector with sample and reference cells and sources out to left and right. The effect is, however, similar to the more usual parallel configuration. The majority of manufacturers now provide for a flowing reference if required, so most analysers can be used either as absolute or differential instruments. However, only the ADC 225 is internally configured so that the same instrument can readily be used as either. The Binos II has provision for the addition of a second optical bench so that two analysers can be contained within the same half rack width box, e.g. absolute and differential for CO_2, or CO_2 and H_2O. In several makes of analyser more than one Luft detector can be put in the optical pathways so that more than one component gas can be analysed, thus allowing cross-sensitivity to be evaluated (e.g. Horiba VIA 300, 500).

The length of the reference and sample cells can be changed in nearly all analysers to enable the same basic instrument to be used in μmol mol^{-1} and percentage ranges. With a particular length of cell, switching over several ranges, usually two to five, is common with linearisation provided on two or three of the ranges.

In most analysers the reference and sample cells are made to be highly reflective internally, using expensive techniques such as gold plating onto glass, to maximise radiation transmittance. In some recent instruments, however, this requirement has been relaxed through the use of imaging optics (e.g. ADC RF, IRI series 700. Anatek 402, Anarad APPA series).

Commercial single beam-path analysers are becoming more common. The Liston Edwards has a single beam path but two heating filaments as mentioned earlier. Radiation from one heating filament passes through the sample cell to a Luft detector; radiation from the other filament passes through both a reference cell, containing the same species of gas, and the sample cell to the detector. Much the same effect is achieved in the ADC RF analyser by the two sealed gas cells in the rotating chopper disc (Fig. 3.5b). In both analysers the gas in the sealed reference cell absorbs considerably more radiation than the sample, thus providing a comparatively stable pulse height against which absorption by the sample alone is compared. A major advantage of these and other single beam-path analysers is that other gas and dirt in the sample path affect both pulse heights to the same degree so that these sources of error are minimised. There is a trend towards making such analysers pseudo-differential by alternating the flow of reference and sample gases and electronically storing and subtracting the signals. This may, however, substantially increase the response time to, say, 10 s.

Laboratory application. There is little to choose between the analysers in Table 3.2 for straightforward use in laboratory experiments in a normally heated/air-conditioned building free from persistent vibration. For monitoring the CO_2 content of the atmosphere, exceptionally good resolution and high stability in the 250 to 500 μmol mol^{-1} range is required, and this is met by the specification of several instruments. For measurements of photosynthesis and respiration in the laboratory, major considerations are also likely to be availability of differential operation, convenience of use and calibration, price, back-up service and import restrictions.

Field laboratory application. A much higher degree of environmental tolerance is desirable in analysers to be used in field laboratories, trailers and vehicles than in normal laboratories. Since power is often supplied from local generators, tolerance of frequency fluctuations is all important, and in addition higher tolerances of vibration and of variations in temperature and humidity are necessary than in the normal laboratory. Several instruments with this combination of features are evident in Tables 3.2 and 3.3.

Portable field systems. We take a portable system to be one that is "convenient for carrying" like, for example, a portable radio (Concise Oxford Dictionary). Analysers for use in portable systems should not only be tolerant of vibration and wide fluctuations in temperature and humidity, but should also be light-weight and operable from battery supplies, without the use of a d.c. to a.c. convertor. There are several fairly coarse instruments available that are adequate for the measurement of animal respiration in the percentage CO_2 range, but for measurement of photosynthesis at normal ambient CO_2 concentrations the choice is very limited (Table 3.3). The Binos II is used in two portable systems (Griffiths and Jarvis, 1981; Schulze *et al.*, 1982) and is, as far as we know, the only commercial battery powered differential analyser available. There are several portable absolute analysers, including the ADC RF and LCA, the Anatek 404 and the Liston Edwards which has been adopted by Li-Cor, Inc., for use in their LI-6000 Portable Photosynthesis System.

A home-built portable differential analyser for use with leaf chambers on cereals is described by Williams *et al.* (1982). A version of the open path analyser developed by Bingham *et al.* (1978, see below) has also been used with leaf chambers (Bingham *et al.*, 1981).

Eddy correlation specials

Measurement of CO_2-flux by eddy correlation depends on measuring the naturally occurring, rapid (0.02 to 2 Hz) fluctuations in CO_2 concentration at a point. Since ducting a parcel of air through the sample cell of a conventional infra-red gas analyser destroys the structure of the air parcel, none of the analysers so far described could be regarded as ideal for eddy correlation. Nevertheless, by pushing air very fast (500 $cm^3\ s^{-1}$) through a closed path infra-red gas analyser in the field, it is possible to measure fluctuations in CO_2 concentration at a point, that closely follow the fluctuations in temperature measured with a rapid response thermocouple *in situ* (Desjardins and Lemon, 1974; Denmead, 1984). Consequently, if some allowance is made for the time delay and for the loss of information at the higher frequencies, it is just possible to use one of the faster conventional analysers for eddy correlation. This suggestion is, however, marginal and the future of CO_2-flux measurement by eddy correlation must lie in the development of suitable small, rapid, open-path CO_2 analysers that can measure fluctuations of CO_2 concentrations *in situ*. Several such instruments have been constructed in the last ten years and used for measuring CO_2-fluxes over vegetation of different kinds (Ohtaki and Seo, 1976 a and b; Ohtaki, 1980; Ohtaki and Matsui, 1982; Brach *et al.*, 1981, Desjardins *et al.*, 1982) and over the ocean (Jones *et al.*, 1978). The specifications of four recent models are listed in Table 3.4. The instrument developed at the Lawrence Livermore Laboratory by Bingham *et al.*, (1978, 1983) has been used for eddy correlation on towers over vegetation and on aircraft over sea and ice.

All four instruments have the same common components as the commercial analysers discussed earlier, i.e. source, chopper, filters and detector, but differ in that

Table 3.4 Properties of four open-path instruments for measuring CO_2 fluxes by eddy correlation.

	Bingham[1]	Ohtaki[2]	Brach[3]	Jones[4]
Beam path	single	single	single	dual
Length of absorbing path (mm)	500	200 or 300	1500	(100?)
Path folded	x12	–	x2	–
Imaging optics	+	+	+	+
Source	black body	nichrome	nichrome	black body (x2)
Source temperature (K)	873	1000	1100	?
Detector	pyroelectric	PbSe	PbSe	PbSe (x2)
Detector temperature (°C)	ambient	–10	–20	cooled
Chopper speed (Hz)	10	30	960	1000
Chopping action	change filters	change filters	± source	ref./sample beams
Filter oscillating speed (Hz)	n.a.	n.a.	10	n.a.
Filter transmission peak (μm)	3.8,4.3	2.6,3.9,4.3	3.7,4.3	4.3
Gas filters	–	–	–	+
Optics purged	+	+	–	+
Resolution (μmol mol^{-1})	0.25	0.5	0.3	0.3
Response time (ms)	100	33	75	100
Microprocessor	1802	Z80	–	–
Battery supply	12V d.c.	–	–	–

[1]Bingham *et al.* (1978,1983), [2]Ohtaki and Matsui (1982), [3]Brach *et al.* (1981), [4]Jones *et al.* (1978)

Abbreviations see Table 3.2

CO_2 absorption is measured in an open path without ducting the air into an enlosed tube (Fig. 3.6). Interference from dust and other gases is reduced by comparing the transmission of radiation at 4.3 μm with the transmission at a nearby wavelength (3.7, 3.8 or 3.9 μm) in the three single beam instruments. In two of the three, this is simply achieved by mounting the appropriate interference filters in the chopper wheel as in the closed path Anatek analysers.

The single beam instrument of Ohtaki and Matsui (1982) has an additional filter in the chopper disc. This facilitates simultaneous measurement of water vapour and CO_2 in the same pathway thus enabling direct comparison of H_2O and CO_2 eddy fluxes. The instrument by Brach *et al.* (1981) differs in that the filters are alternately oscillated in and out of the beam path at 10 Hz, whereas the radiation source is separately chopped at 960 Hz.

The dual beam instrument of Jones *et al.* (1978) resembles conventional closed-path dual beam absolute analysers in several respects. Beams of radiation from separate

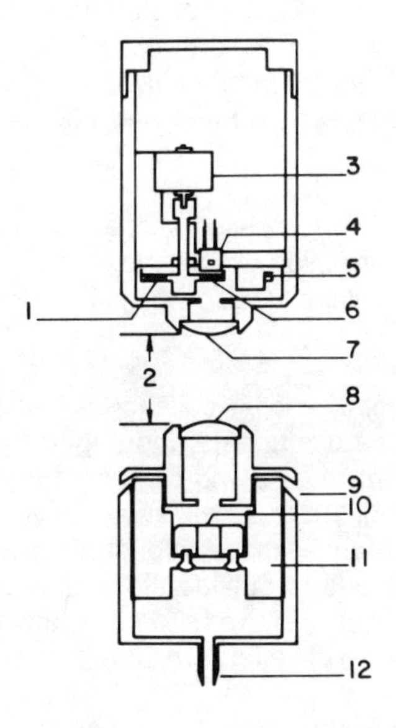

1. 2.6 μm filter
2. Measurement path 20cm
3. Pulse motor
4. PbSe detector
5. Photo sensor
6. 4.3 μm filter
7. CaF_2 lens
8. CaF_2 lens
9. Air inlet
10. Infra-red source
11. Fin
12. Air outlet

Figure 3.6 A diagram of the open path analyser developed for eddy correlation measurements by Ohtaki and Matsui (1982).

reference and sample sources are alternately chopped after the sample beam has traversed the open path. The reference path is closed and contains zero CO_2. After chopping, the beam paths are combined and pass through a single interference filter (4.3 μm peak transmission) before being split again to the two separate detectors, one beam passing through a CO_2-containing gas filter, the other a neutral gas filter.

All four instruments have solid state detectors, three of them using the same make of lead selenide device. These need to be kept at low temperatures and the consequent requirement for thermo-electric cooling substantially increases power consumption. The Bingham *et al.* (1978) instrument utilises a pyro-electric detector and low power electronics and can consequently be operated from a 12 volt battery. Pyro-electric sensors are, however, sensitive to vibration and need special care (Bingham *et al.*, 1981).

The sample beam path length varies widely amongst the four. The Brach *et al.* (1981) instrument has the longest path of 1.5 m. Source and detector are side by side, the beam being reflected back to the detector by a mirror after a traverse of 0.75 m. The sensor developed by Bingham *et al.* (1978) makes approximately 16 traverses over a distance of only about 30 mm to give a 0.5 m pathlength (i.e. it is a folded path sensor). In the other two instruments the beam makes a single traverse from source to detector over a distance of between 100 mm and 500 mm, depending on their relative positions.

Two of these instruments include microprocessors. These facilitate on-line calculation and allow the ready inclusion of correction factors, such as the effects of temperature and pressure on the sample gas, linearisation and the influence of temperature on filter band pass (Bingham *et al.*, 1978).

The instruments developed by Ohtaki and his colleagues have been used in the field since the early 1970s and provide useful data about the rapid fluctuations of CO_2 density over crops. Power spectra of CO_2 and cospectra of CO_2 and vertical wind speed over rice and wheat (Ohtaki, 1980; Ohtaki and Matsui, 1982), are very similar to published work for water vapour and demonstrate that there is negligible contribution to the CO_2-flux from eddies outside the frequency range 0.02 to 2 Hz. Consequently, the four instruments referred to here should all be capable of measuring CO_2-flux with an accuracy of better than 10%.

Calibration of infra-red analysers

Infra-red gas analysers need to be calibrated empirically for both zero and sensitivity and the calibration should be repeated frequently, the frequency depending on the design, measurement range and required accuracy of measurement. The most obvious reasons for changes in calibration are changes in temperature and pressure of the measured gas, ageing of the source and detector, changes of wall reflectivity, window transmittance in the optical paths, and, in closed-path instruments particularly, the entry of dirt. Also with a differential analyser, the zero output (i.e. when $\rho_r = \rho_s$) changes with changes in ρ_r because the absorptance properties of the two optical paths are rarely identical. This zero drift is in addition to the change in sensitivity that results from change in ρ_r (see "Principle of infra-red gas analyser" earlier in this chapter). Thus differential analysers require more frequent calibration than absolute analysers.

In the majority of analysers there is no provision for calibration other than the introduction of gases of known density. In a number of analysers a rough check can be obtained by insertion of a wire or a gas containing cell in the optical path. More sophisticated internal, automatic recalibration systems are incorporated in the Binos II, as described earlier, and in the IRI series 700.

Calibration principles. The purpose of calibration is to scale the output of the analyser, perceived as an electrical signal or meter reading, into the quantities and units required for subsequent calculations. Infra-red gas analysers measure the partial density of CO_2 in air (ρ, kg m^{-3}) or partial pressure (Pa) – see equation 3.1; they do *not* measure volume fraction (c, m^3 m^{-3}), mole fraction (x, mole $mole^{-1}$) or mixing ratio (kg kg^{-1}). Calibration is done by passing scale-point gases of known CO_2 content through the analyser to obtain outputs that depend on the CO_2 *density* of the gases. However, the CO_2 content of scale-point gases is usually defined in terms of *volume* or *mole fraction* because scale-point gases are often prepared by volumetric or mass-mixing. These quantities have the advantage that they are independent of temperature and pressure, so that correctly determined values of scale-point gases can be readily transferred from time to time and place to place.

The partial CO_2 density is a function of temperature (T,K) and pressure (P,Pa) of the gas and is related to c or x by:

$$\rho = x\rho_o(P/P_o)(T_o/T) = x5.239\,10^{-3}(P/T) \qquad 3.3$$

where the subscript o indicates standard temperature and pressure. Thus the density has a fixed relationship to the mole fraction only at one particular temperature and pressure. If the output is scaled to mole fraction at P_1 and T_1 and an unknown sample is then measured also at P_1 and T_1, the mole fraction of the sample gas is found directly from the scale without the need for any corrections for temperature and pressure. If, however, the temperature or pressure of gas in the analyser changes between calibrating the analyser and measuring the sample, it is necessary either to recalibrate, to compensate automatically for the change or to correct for the change by calculation.

For thermostatted analysers, temperature of the gas can be regarded as constant, but pressure is likely to change as a result of changes in the weather and in gas flow rate. For unthermostatted analysers changes in both T and P are likely. Regular recalibration is the most straightforward solution in many circumstances. Alternatively, automatic compensation for changes in T or P is offered as an option on some analysers (see Tables 3.2 and 3.3) and this is likely to be of particular relevance when analysers are used in the field away from calibration facilities.

Correction for changes in T and P is based on the gas laws. The CO_2 density at T_1 and P_1 is related to that at T_2 and P_2 by:

$$\rho_1 = \rho_2(P_1/P_2)(T_2/T_1) \qquad 3.4$$

Since analyser output is proportional to density, whatever units the output is expressed in, equation 3.4 can be used directly to correct the output (V) for any differences in temperature between the sample gas and the scale-point gases. However, equation 3.4 is not wholly adequate to correct for similar differences in pressure because of the effects of pressure in broadening the radiation absorption lines of CO_2 (Parkinson and Legg, 1971). A corrective factor (z), which relates the response to pressure found in practice to the response predicted by equation 3.4, can be determined empirically (Legg and Parkinson, 1968; Bate *et al.*, 1969) and applied using:

$$V_1 = \frac{V_2 P_1}{z(P_2 - P_1) + P_1}\,\frac{T_2}{T_1} \qquad 3.5$$

z is characteristic of an individual analyser and has been found to vary between 1.1 and 1.7.

Whether to recalibrate regularly, to compensate or to correct is a matter of choice with most closed path analysers. There is no alternative, however, to making this correction in eddy-correlation studies, as variations in temperature and pressure are inherent in the measurements.

In differential analysers the temperature and pressure in the reference and sample cells should also be identical. Whilst this is normally true for temperature (except at very high flow rates), pressure varies with flow rate and resistance to flow in the gas lines and consequently can easily differ between cells. Such a difference in pressure between the cells of only 100 Pa (approximately 10 mm H_2O) results in an apparent CO_2 differential of 0.3 μmol mol^{-1} (at $P = 10^5$ Pa and $x = 300$ μmol mol^{-1}), therefore accurate measurement or control of the pressure differential between the cells, at the point of gas entry, is essential for accurate measurement of the CO_2 differential. Thus a differential analyser should always be used in conjunction with a sensitive differential pressure transducer. If flow rates are very high, temperature equilibration can be ensured by passing the reference and sample gas streams through copper coils within the analyser case.

A further problem is that the composition of commercial scale-point gases is often inadequately defined. Gases purchased for routine recalibration should be compared against standard gases of known provenance and accuracy. Unfortunately there is substantial uncertainty attached to the best primary standards that one can get (Bate *et al.*, 1969). A quoted accuracy of $\pm 5\%$ is not uncommon and $\pm 1\%$ is difficult to obtain. Uncertainty of $\pm 5\%$ on each of a pair of gases renders them completely unsuitable for the calibration of a differential analyser with a difference of, say, 10% of ρ_r. Acceptable absolute accuracy is easier to get using a system in which either volumes (e.g. with Wosthoff pumps) or masses (e.g. with mass flow meters) are mixed precisely. A novel, compact, portable device is now available which mixes pure CO_2 and CO_2 free air by volume (Li-cor, model LI-6000-01). This has an absolute accuracy of $\pm 1.4\%$ with a repeatability of $\pm 0.5\%$, so it is suitable for field calibration of both absolute and differential analysers. Accurate relative differences between gases may be generated using one standard gas and a gas diluter (e.g. ADC GD-300), or a single gas mixing pump. When using cylinders one must also be aware that the composition of the standard gases may change during storage, depending on the type of container and water content of the gas (Janac *et al.*, 1971). Scale-point gases should be dry to minimise the risk of interference from water vapour.

Calibration procedures. To calibrate an *absolute* analyser the simplest procedure is to define that part of the curve shown in Fig. 3.1 that lies in the range of interest, by passing several different scale-point gases through the sample cell. It is important to check the degree of linearity if the instrument has been electrically linearised at the factory, e.g. with a gas diluter: if the response is properly linear within the specification then it is only necessary to use two or three scale-point gases regularly. The "tube-length" method of recalibration (see below) can also be used to obtain two scale points regularly from one known scale-point gas that has a mole fraction in the area of interest. If the analyser has not been linearised but is used only over a restricted part of the response curve, it may be satisfactory to assume linearity over that part of the range, but the error involved should be calculated. If an unlinearised analyser is to be used over the whole of the range, the shape of the curve should be defined and an equation fitted to it.

Because the sensitivity of a *differential* analyser depends on the mole fraction of the flowing reference gas, it is desirable *throughout* calibration to have the same gas in the reference cell as during measurement. A suitable procedure when using an unknown flowing reference gas is to pass the unknown sample, the unknown reference and two known scale-point gases in succession through the sample cell and record the output, whilst the same unknown reference gas continues to flow through the reference cell (Table 3.5). Then the sensitivity of the instrument (S) is given by $S = (V_4 - V_3)/(x_2 - x_1)$. The flowing reference mole fraction is $x_r = x_1 + (V_2 - V_3)/S$ and the sample mole fraction is $x_s = x_r - (V_2 - V_1)/S$.

Table 3.5 A calibration procedure for differential analysers using two or more scale-point gases.

Operation	Gas in		Output
	reference cell	sample cell	
measurement	reference x_r	sample x_s	V_1
zero	reference x_r	reference x_r	V_2
calibration 1	reference x_r	scale-point x_1	V_3
calibration 2	reference x_r	scale-point x_2	V_4

This procedure assumes linearity in the region of interest so if the analyser is unlinearised the differential $(x_r - x_s)$ should be kept small and the scale-point gases should narrowly bracket the reference and sample gases.

A similar procedure is used in the so-called "tube-length" method (Legg and Parkinson, 1968; Parkinson and Legg, 1971, 1978). In this method the sample cell is divided into a long cell and a short cell with an appropriate ratio of tube lengths [$l_2/(l_1 + l_2) = \Phi$, Fig. 3.7]. By passing CO_2-free air through the short cell, an output is obtained that corresponds to a CO_2 differential over the total tube length of Φx_r, if x_r is the mole fraction of gas in the reference cell and the long sample cell. A suitable procedure is given in Table 3.6. The same assumption of linearity is made as before. Calculations made using a hyperbolic function fitted to response curves as in Fig. 3.1 show that this assumption results in less than 1% error (with Φ of 5% at x_r of 300 μmol mol^{-1}). Using the same notation as before, $S = (V_4 - V_3)/(\Phi x_1)$, $x_r = (V_2 - V_3)/(\Phi S)$, and $x_s = x_r - (V_2 - V_1)/S$.

The value of Φ depends on the application: about 5% is appropriate for measurements with assimilation chambers and 1 or 2% for measurements of atmospheric profiles. Φ should be known to the required accuracy of the CO_2 measurement; it can be determined by measurement of the tube lengths but should be checked using standard gases. This method has the great advantage that in practice only one, accurately known, scale-point gas is needed in any range. The method is particularly suitable if known flowing reference gases over a wide range of CO_2 mole fractions are being supplied to a chamber (e.g. from mixing pumps or mass flow controllers) since no additional scale-point gases are then needed and the fourth step can be omitted. Other advantages are listed by Parkinson and Legg (1978). The ADC 225 is the only commercial analyser configured for routine recalibration using this method. However, by specifying cells of appropriate lengths and fitting additional ports for gas lines, the method can easily be fitted to many instruments (e.g. URAS 3G).

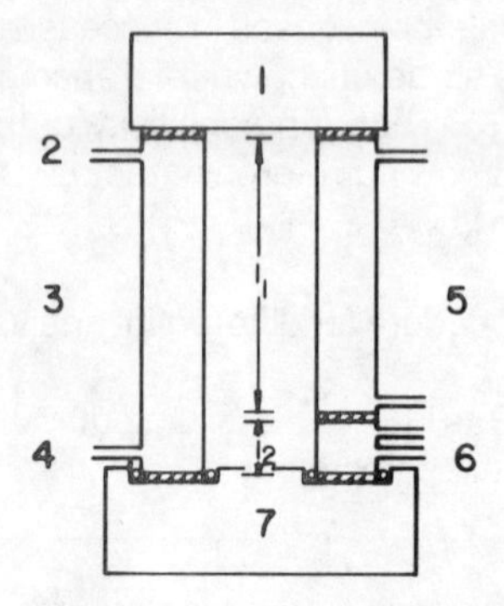

Figure 3.7 A diagram to show the configuration of reference and sample cells for the "tube length" calibration method. l_1 is the length of the long sample cell; l_2 is the length of the short sample cell (from Parkinson and Legg, 1978).

Thorpe (1978) and Bloome *et al.*, (1980) have described procedures for formalising the dependence of S on x_r over a wide range of values of x_r by fitting their calibration data to hyperbolic functions. It is noteworthy, however, that Thorpe's equations reduce to those given above, provided that the same gas is present in the reference cell throughout each stage of calibration and during measurement, as in Table 3.6. Their approach may be useful in defining the whole response surface in some circumstances, but it cannot, for example, take changing temperature and pressure into account. In our view, for the accurate measurement of small differentials or small variations of ambient mole fraction, it is essential to recalibrate each time measurements are made. This is readily done routinely at preset time intervals or once per cycle of measurements by programmed, automatic switching of gases with solenoid valves through one of the procedures outlined above.

Calculation of fluxes

The calculation of fluxes of CO_2 is extensively covered in the literature. However, several recent theoretical studies have revealed oversights in the earlier theory and shown that errors arise as the result of a simultaneous flux of water vapour and/or heat. As correction of these errors may affect the method of CO_2 measurement, they are considered here.

Firstly a correction is necessary to allow for variation in the water content of gas samples measured by the analyser. The infra-red gas analyser correctly and simply measures the CO_2 density or the mole fraction, if any changes in T and P are allowed

Table 3.6 A calibration procedure for differential analysers using the tube length method.

Operation	Gas in			Output
	reference cell	sample cell		
		long	short	
measurement	reference x_r	sample x_s	sample x_s	V_1
zero	reference x_r	reference x_r	reference x_r	V_2
calibration 1	reference x_r	reference x_r	CO_2-free	V_3
calibration 2	reference x_r	reference x_r	scale-point x_1	V_4

for. However, the mole fraction of CO_2 in a sample also depends on the molar fractions of the other constituents. If the mole fraction of water vapour of a sample is increased by transpiration, for example, from x_{W1} to x_{W2}, without the loss or gain of CO_2, the mole fraction of CO_2 will, nonetheless, be decreased from x_1 to x_2. If the two samples are passed through a differential gas analyser there will be an output that is proportional to $(x_1 - x_2)$ (Parkinson, 1971). Correction can be made for this according to:

$$x_1 = x_2((1-x_{W2})/(1-x_{W1})) \qquad 3.6$$

The gas streams should either be dried before entering the analyser or this correction should be made in both assimilation chamber studies (Parkinson, 1971; von Caemmerer and Farquhar, 1981; Leuning, 1983), and in mean gradient studies (Webb *et al.*, 1980). If cross-sensitivity to water vapour has not been completely eliminated an additional empirical correction is necessary. Before the use of thin-film filters became widespread to reduce cross-sensitivity to H_2O, it was usual to dry both reference and sample gas streams before the analyser (Jarvis *et al.*, 1971). This practice obviates the need for both corrections and is the only answer when x_W is not measured. In open path systems drying the air in the pathway is impossible and this correction must, therefore, be made in eddy correlation studies (Webb *et al.*, 1980; Leuning *et al.*, 1982).

Secondly, a correction is required in the calculation of fluxes of CO_2, in the presence of fluxes of heat and water vapour. Consider a flux of water vapour from Z_1 to Z_2 in the absence of any net flux of CO_2 e.g. through the stomata at the CO_2 compensation point, or in the air above bare soil. The flux of water vapour will be accompanied by a gradient in water vapour density between Z_1 and Z_2 and this must be accompanied by a gradient of "dry air" (including CO_2) in the reverse direction. There will, however, be no net flux of dry air since there is no sink for dry air in the leaf or at soil surface. Thus there will be a gradient of CO_2 in the absence of any CO_2 flux. This applies both in the case of a leaf (Jarman, 1974; Leuning, 1983) and the atmosphere (Webb *et al.*, 1980). Failure to take this into account can lead to errors of 5% in the calculated mean mole fraction in the leaf (Jarman, 1974; Leuning, 1983) and 90% in the CO_2 flux measured by eddy correlation (Leuning *et al.*, 1982).

CO_2 Measurement in the future

Future developments are liable to depend very largely on industrial requirements. The large, immobile, dual beam instruments have been, and will probably continue to be rather conservative to changes in design, although reduced sensitivity to mains fluctuations, vibration and interfering gases has been achieved in different ways in recent instruments. A combination of these recent innovations will probably result in a trend towards a single source, dual beam configuration (to allow differential measurement) with an advanced high speed mechanical chopper and solid-state detector. Improved specificity of the detector may be achieved by using a mass flow sensor with two CO_2-filled absorption chambers arranged in series. In addition, either thin-film filters, or extra, species-specific, solid state detectors may be used to reduce cross-sensitivity to water vapour. These instruments are likely to remain the first choice for laboratory measurement of CO_2, particularly when measuring small CO_2 differentials.

More recent requirements of industry have included rugged instruments for pollution monitoring, industrial horticulture and deep-sea diving. Manufacturers have developed new instruments to meet these demands, many of which are based on a single beam configuration, providing absolute measurement only. These designs are likely to be improved by new solid state choppers and solid state sensors. Solid state, pulsating sources also reduce sensitivity to vibration but introduce long term drift as two sources are required. Advances in liquid-crystal technology may offer an alternative means of solid state chopping (Bingham *et al.*, 1978).

The portability and comparative ruggedness of these new instruments make them ideal for measuring CO_2 in the field. One at least has been modified for open path measurement, and has possibilities for use in eddy correlation studies. Others may be suitable for portable, field photosynthesis systems: both transit-time (e.g. Li-Cor., LI6000) and null-balance techniques (Griffiths and Jarvis, 1981) make use of absolute measurement of CO_2. For both techniques, a folded path, miniature CO_2 sensor mounted in the cuvette would be ideal (Bingham *et al.*, 1980, 1981). Future designs of portable, single beam instruments may well fill this requirement. Lasers are already being used for infra-red spectroscopy of hydrocarbons, and it can only be a short time before they are applied to CO_2. Because they produce monochromatic, narrow-beam radiation (see Sheehy, Chapter 2), many of the problems of conventional instruments would be avoided. Recent advances in electronics will also lead to improvements in analyser performance. The incorporation of digital processing of the analyser output, and even some inbuilt logging capability may well appear as an option for both the laboratory and field instruments.

We have been surprised and impressed by the large number of innovations distributed amongst the instruments described here, especially the eddy correlation specials. We anticipate that manufacturers will continue to utilise these innovations in the design of new analysers that will be suitable for our applications.

Manufacturers

Infra-red gas analysers

Where applicable, the address of the parent company is given in parentheses.

The Analytical Development Company Ltd., (ADC) Pindar Road, Hoddesdon, Hertfordshire EN11 0AQ, England.

Anarad, Inc., P.O. Box 3160, Santa Barbara, CA93105, U.S.A.

Anatek Instruments Ltd., High Street, Mayfield, Sussex TN20 6AB, England.

Beckman-RIIC Ltd., 6 Stapledon Road, Orton Southgate, Peterbourgh PE2 0TB, England. (Beckman Instruments Inc., 2500 Harbor Boulevarb, Fullerton, CA 92634, U.S.A.)

Constructions et services pour la mesure et l'analyse (COSMA). Chemin de la Sabliere, Zone Industrial, 91430 IGNY, France.

Edinburgh Instruments Ltd., Riccarton, Currie, Edinburgh EH14 4AP, Scotland.

Hartmann and Braun Ltd., (H and B), Moulton Park, Northampton NN3 1TF, England.(Hartmann and Braun AG, Postfach 90 0507, GrafstraBe 97, D-6000 Frankfurt 900, F.D.R.)

Horiba Instruments Ltd., 5 Harrowden Road, Brackmills, Northampton NN4 0EB, England. (Horiba Ltd., Miyanohigashi, Kisshoin, Minami-ku, Kyoto, Japan.)

Infra-red Industries (IRI), from Auriema Ltd., 442 Bath Road, Slough, Buckinghamshire SL1 6BB, England. (Infra-red Industries, P.O. Box 989, Santa Barbara, CA 93102, U.S.A.)

Leybold-Heraeus Ltd., 16 Endeavour Way, Durnsford Road, London SW19 8UH, England. (Leybold-Heraeus GMBH, Wilhelm-Rohn-StraBe 25, P.O.B. 1555, D-6450 Hanau 1, F.D.R)

Liston Edwards Inc., 3800 Campus Drive, Newport Beach, CA 92660, U.S.A.

H.Maihak AG, from Smail Sons and Co. Ltd., 129 Whitfield Road, Glasgow G51 2SF, Scotland. (H.Maihak AG, Semperstrasse 38, 2000 Hamburg 60, F.D.R.)

Miran, from Quantiteck Ltd., 75 Garamonde Drive, Wymbush, Milton Keynes, Buckinghamshire MK8 8DD, England. (Foxboro Analytical, P.O. Box 5449, 140 Water Street, South Norwalk, CT 06856, U.S.A.)

J. and S. Sieger Ltd., 31 Nuffield Estate, Poole, Dorset BH17 7RZ, England.

Accessories

Gas mixtures for calibration

Air Products Ltd., Special Products Dept., Weston Road, Crewe CW1 1DF, England.

The British Oxygen Company (B.O.C.), Special Gases, Deer Park Road, London SW19 3UF, England.

P.K.Morgan Ltd., 4 Bloors Lane, Rainham, Kent ME8 7ED, England.

Rank Hilger Ltd., Westwood, Margate, Kent CT9 4JL, England.

Tubing and fittings:

Cole-Parmer, P.O. Box 22, Bishop's Stortford. Hertfordshire, England. (Cole-Parmer International, 7425 North Oak Park Avenue., Chicago, IL 60648, U.S.A.)

Samuel Moore Europe S.A., Unit 16, Midland Oak Industrial Estate, Lythalls Lane, Coventry CV6 6FJ, England.

Gas blending systems:

A.D.C. Ltd., (see IRGA manufacturers).

B.O.C., Special Gases (see suppliers of gas mixtures).

Brookes Instrument Div., Brooksmeter House, Stuart Road, Bredbury, Stockport SK6 2SR, England. (Emerson Electric Co., Hatfield, PA 19440, U.S.A.)

Horiba Instruments Ltd., (see IRGA manufaturers).

Hastings, from Chell Instruments Ltd., Tudor House, Grammer School Road, North Walsham, Norfolk NR26 9JH, England. (Teledyne Hastings-Raydist, P.O. Box 1275, Hampton, VA 23611, U.S.A.).

LI-COR, inc., P.O. Box 4425, 4421 Superior St., Lincoln, Nebraska 68054, U.S.A.

Signal Instrument Co. Ltd., St.Mary's Works, Krooner Road, Camberly, Surrey GU15 2QP, England.

Tylan, from Epak Electronics Ltd., Pool House, Bancroft Road, Reigate, Surrey RH2 7AP. (Tylan Corp. 23301 SO. Wilmington Avenue., Carson, CA 90745, U.S.A.)

H. Wosthoff, from Kandem Electrical Ltd., 711 Fulham Road, London SW6 5UN, England. (H. Wosthoff OHG, D-4630 Bochum 1, Max-Greve-StraBe 30, F.D.R.)

Sampling pumps:

Charles Austen Pumps Ltd., 100 Royston Road, Byfleet, Weybridge, Surrey KT14 7PB, England.

Brailsford and Co. Inc., Milton Road, Rye, NY10580, U.S.A.

Brey, Dipl. Ing. Helmut Brey, LuitpoldstraBe 28, Postfach 1329, 8940 Memmingen, F.D.R.

KNF Neuberger U.K. Ltd., KNF House, Ardington, Wantage, OX12 8QZ, England. (KNG Neuberger GmbH, D-7800 Freiburg-Munzingen, F.D.R.)

References

Bate, G.C., D'Aoust, A. & Canvin, D.T. (1969). Calibration of infra-red CO_2 gas analysers. Pl. Physiol., **44**, 1122-1126.

Bingham,G.E., Coyne, P.I., Kennedy, R.B. & Jackson, W.L. (1980). Design and fabrication of a minicuvette for measuring leaf photosnthesis and stomatal conductance under controlled conditions. Lawrence Livermore Laboratory Rept. UCRL-52895.

Bingham, G.E., Gillespie, C.H. & McQuaid, J.H. (1978). Development of a miniature, rapid response carbon dioxide sensor. Lawrence Livermore Laboratory Rept., UCRL-52440.

Bingham, G.E., Gillespie, C.H., McQuaid, J.H. & Dooley, D.F. (1981). A miniature, battery powered, pyroelectric detector-based differential infra-red absorption sensor for ambient concentrations of carbon dioxide. Ferroelectrics, **34**, 15-19.

Bingham, G.E., Gilmer, R.D., Maish, F.M. & Hansen, K. (1983). Area CO_2 flux studies: Aircraft measurements over open ocean and sea ice. Annual Report to the U.S. Dept. of Energy, CO_2 and climate office (in press).

Bloom, A.J., Mooney, H.A., Bjorkman, O. & Berry, J. (1980). Materials and methods for carbon dioxide and water exchange analysis. Plant, Cell & Environ., **3**, 371-376.

Brach, E.J., Desjardins, R.L. & StAmour, G.T. (1981). Open path carbon dioxide analyser. J. Phys., E. (Scientific Instruments), **14**, 1415-1419.

Coombes, R.G. & Stroud, D.J. (1982). A new concept in solid state infra-red gas analysers. International Environment & Safety, June 1982.

Denmead, O.T. (1984). Plant physiological methods for studying evapotranspiration: problems in telling the forest from the trees. Agricultural Water Management, **8**, 167-189.

Desjardins, R.L. & Lemon, E.R. (1974). Limitations of an eddy-correlation technique for the determination of the carbon dioxide and sensible heat fluxes. Boundary-Layer Meteorol., **5**, 475-488.

Desjardins, R.L., Brach, E.J., Alvo, P. & Schuepp, P.H. (1982). Aircraft monitoring of surface carbon dioxide exchange. Science, N.Y., **216**, 733-735.

Griffiths, J.H. & Jarvis, P.G. (1981). A null-balance carbon dioxide and water vapour porometer. J. exp. Bot., **32**, 1157-1168.

Janac, J. (1970). The accuracy of differential measurements of small CO_2 concentration differences with the infra-red analyser. Photosynthetica, **4**, 302-308.

Janac, J., Catsky, J., Jarvis, P.G. *et al.* (1971). Infra-red gas analysers and other physical analysers. *In* Plant Photosynthetic Production: Manual of Methods, eds. Z. Sestak, J. Catsky and P.G. Jarvis, pp. 111-197. The Hague: Dr W. Junk.

Jarman, P.D. (1974). The diffusion of carbon dioxide and water vapour through stomata. J. exp. Bot., **25**, 927-936.

Jarvis, P.G., Catsky, J., *et al.* (1971). General principles of gasometric methods and the main methods of installation design. *In* Plant Photosynthetic Production: Manual of Methods, eds. Z. Sestak, J. Catsky and P.G. Jarvis, pp. 49-110. The Hague: Dr W. Junk.

Jones, E.P., Ward, T.V. & Zwick, H.H. (1978). A fast response atmospheric CO_2 sensor for eddy correlation flux measurements. Atmos. Environ., **12**, 845-851.

Keeling, C.D., Bacastow, R.B., Bainbridge, A.E., Ekdahl, C.A., Guenther, P.R & Waterman, L.S. (1976). Atmospheric carbon dioxide variations at Mauna Loa Observatory, Hawaii. Tellus, **28**, 538-551.

Legg, B.J. & Parkinson, K. J. (1968). Calibration of infra-red gas analysers for use with carbon dioxide. J. Phys., E. (Scientific Instruments), **1**, 1003-1006.

Leuning, R. (1983). Diffusion of gases into leaves. Plant, Cell & Environ., **6**, 181-194.

Leuning, R., Denmead, O.T., Lang, A.R.G. & Ohtaki, E. (1982). Effects of heat and water vapour transport on eddy covariance measurement of CO_2 fluxes. Boundary-Layer Meteorol., **23**, 209-222.

Ohtaki, E. (1980). Turbulent transport of carbon dioxide over a paddy field. Boundary-Layer Meteorol., **19**, 315-336.

Ohtaki, E. & Matsui, M. (1982). Infra-red device for simultaneous measurement of atmospheric carbon dioxide and water vapour. Boundary-Layer Meteorol., **24**, 109-119.

Ohtaki, E. & Seo, T. (1976a). Infra-red device for measurement of carbon dioxide fluctuations under field conditions. II. Double beam system. Ber. Ohara Inst. landw. Biol. Okayama Univ., **16**, 175-182.

Ohtaki, E. & Seo, T. (1976b). Infra-red device for measurement of carbon dioxide fluctuations under field conditions. II. Double beam system. Ber. Ohara Inst. landw. Biol. Okayama Univ., **16**, 183-190.

Parkinson, K.J. (1971). Carbon dioxide infra-red analysis: effects of water vapour. J. exp. Bot., **22**, 169-176.

Parkinson, K.J. & Legg, B.J. (1971). A new method for calibrating infra-red gas analysers. J. Phys., E. (Scientific Instruments), **4**, 598-600.

Parkinson, K.J. & Legg, B.J. (1978). Calibrations of infra-red analysers for carbon dioxide. Photosynthetica, **12**, 65-67.

Schulze, E.D., Hall, A.E., Lange, O.L. & Walz, H. (1982). A portable steady-state porometer for measuring the carbon dioxide and water vapour exchanges of leaves under natural conditions. Oecologia, **53**, 141-145.

Schunck, G. (1978). Non-dispersive infra-red gas analysers for industrial processes and the protection of the environment. Meas. & Control, **11**, 245-247.

Sestak, Z., Catsky, J. & Jarvis, P.G. (1971). Plant Photosynthetic Production: Manual of Methods. The Hague: Dr W. Junk.

Thorpe, M.R. (1978). Correction of infra-red gas analyser readings for changes in reference tube CO_2 concentration. Plant, Cell & Environ., **1**, 59-60.

von Caemmerer, S. & Farquhar, G.D. (1981). Some relationships between biochemistry of photosynthesis and the gas exchange of leaves. Planta, **153**, 376-387.

Webb, E.K., Pearman, G.I. & Leuning, R. (1980). Correction of flux measurements for the density effects due to heat and water vapour transfer. Q. Jl. R. met. Soc., **106**, 85-100.

Williams, B.A., Gurner, P.J. & Austin, R.B. (1982). A new infra-red gas analyser and portable photosynthesis meter. Photosynthesis Res., **3**, 141-151.

Chapter 4: Water vapour measurement and control

W. Day
Long Ashton Research Station, University of Bristol,
Long Ashton, Bristol, BS18 9AF, England.

Introduction

The measurement or control of water vapour concentration in air is important in many areas of environmental physiology but it is difficult to do with accuracy, and many possible methods exist. Before considering these methods and their capabilities, I will review briefly the roles of water vapour in environmental physiology to identify the factors that will determine the appropriateness of a particular method. Some general roles are outlined below: others are considered in detail elsewhere in this volume (porometry, Chapter 10, and plant water relations, Chapter 11).

In many studies it is necessary to "define the physical environment". This applies to many uses of meteorological data and to measurements in glasshouses and controlled-environment chambers. In the latter there is often the additional requirement to regulate the water vapour concentration. However, for both control and measurement, the accuracy required is often low, and simple measurement and recording are sufficient.

The widest use of accurate measurements of absolute water vapour concentration is in the determination of water fluxes e.g. from field crops or from leaves in assimilation chambers or from animals. In general, measurements of water vapour concentration and gas flow rate are required, and it is often the difference between two concentrations that is measured e.g. between input and output gas flow for a chamber or between two heights above a crop canopy.

The most precise determinations of relative humidity are required in studies of water potential (or water activity) of solid or liquid materials. This determination requires that a thermodynamic equilibrium is established between the water vapour in the air and the water in the solid or liquid. For much living tissue this equilibrium humidity is within a few per cent of saturation, and hence high precision is needed to distinguish differences in tissue water potential. In dormant tissues, e.g. seeds, and when considering some microbiological activity, e.g. food storage, much lower equilibrium humidities occur.

As well as these specific interests in water vapour, it must also be measured if its presence interferes with some other measurement, e.g. of CO_2 exchange rates of plants. Transpiration increases the water vapour content of the air in the measurement system, and hence also the volume of the air at constant pressure. This increase in volume decreases the CO_2 concentration independent of any CO_2 exchange. In addition, the CO_2 analyser may be sensitive to water vapour particularly as the changes in CO_2 concentration are often 10^4 times smaller than those of water vapour (Parkinson, 1971).

The range of water vapour concentrations that are of interest to environmental physiologists is wide; from saturation down to 5 to 10% of saturation. The measurement of lower concentrations, for which there are a number of specialised instruments, is rarely required, though dry gas streams are often used, and therefore the efficiency of drying agents is important. The ability to generate known water vapour concentrations throughout the measurement range is also necessary, both to provide suitable environmental conditions for experiments and to calibrate humidity sensors.

Definitions and physical principles

Water vapour in air acts, for all biological purposes, like an ideal gas. Total air pressure is the sum of the partial pressures of the constituent gases, including the water vapour pressure, e, (Dalton's law) and the effect of temperature on the vapour pressure of unsaturated air is given by

$$\rho = e/(CRT) \qquad 4.1$$

where ρ is the vapour density, R is the gas constant, T the absolute temperature and C is the compressibilty factor; C is within 0.3% of unity, the ideal gas value.

The water vapour content of air can be expressed in a number of ways, of which the most important are:

vapour pressure – the partial pressure of water vapour in the air: e, unit Pa;

vapour concentration or vapour density or absolute humidity – the ratio of the mass of water vapour to the volume occupied by the moist air: χ, unit kg m^{-3};

relative humidity – the ratio of actual vapour pressure to the saturation vapour pressure at air temperature: h, dimensionless, but often expressed as a percentage;

specific humidity – the ratio of the mass of water vapour to the total mass of moist air: q, unit g(H_2O) kg^{-1}.

The concept of saturation must be introduced because at temperatures and pressures appropriate to biological systems, water can co-exist as vapour and as liquid water or ice. Air is saturated when its composition is unchanging in the presence of a plane surface of the condensed surface (water or ice) at the same temperature and pressure as the air. The *saturation vapour pressure*, e_s, is the vapour pressure of water in that air. If the vapour pressure exceeds e_s, water will condense out to reduce e to e_s; vapour pressures below e_s are stable unless water is present in the condensed phase, when e will rise towards e_s as the condensed phase evaporates.

The *dew-point temperature* is the temperature to which moist air must be cooled to become saturated with water vapour, at constant pressure and specific humidity.

When the dew-point temperature is below 0°C there are two values for both saturation vapour pressure and dew-point temperature depending on whether the saturated vapour is in equilibrium with water or ice. Though ice is the stable condensed phase below 0°C, water can be cooled below 0°C for long periods without freezing (the majority of clouds at temperatures below 0°C consist entirely or mainly of liquid water). Thus it is common practice always to refer to dew-point temperature (rather than frost-point temperature) and to express relative humidity on the basis of saturation with respect to water.

The *saturation vapour pressure deficit* or just *saturation deficit* is the difference between the saturation and actual vapour pressures of the air. It is commonly used because of its importance in determining evaporation rates. In plant studies, the term *vapour pressure deficit* has been commonly used (see Rawson, Begg and Woodward, 1977; Day, Lawlor and Legg, 1981) for the difference between the vapour pressure in the inter-cellular spaces of leaves (assumed to be saturated at leaf temperature) and that in the air surrounding the leaves.

When air is in equilibrium with materials that contain water, the equilibrium humidity is determined by the energy status of the bound water. The *water potential*, ψ, is the total specific free energy of water in that material (Slavik, 1974), and is related to the relative humidity, h, of the air at equilibrium by

$$\psi = (RT/\overline{V}_w)\log_e h \qquad 4.2$$

where $\overline{V}_w$ is the partial molal volume of water. The value of the equilibrium relative humidity is also referred to as the *water activity* of the material.

Further details of the physical quantities and their definitions relevant to water vapour content of air are given by Smithsonian Institute (1968), including comprehensive tables for saturation vapour pressure and other thermodynamic properties of moist air. Some examples of humidity expressed in different ways are given in Table 4.1, together with comparisons of errors. There have been some recent small improvements in measurement and specification of thermodynamic properties, including the saturation vapour pressure over water (Wexler, 1976). For many purposes, a simple mathematical formula relating e_s to T is useful, and Buck (1981) gives formulae for equilibrium with water or ice and including non-ideal gas dependence on absolute pressure. For saturation over water at 100 kPa and T°C the formula is

$$e_s = 0.61375\exp(17.502T/(240.97+T))\ \text{kPa} \qquad 4.3$$

Techniques and instrumentation

Many of the techniques available for measuring or controlling water vapour in air have been in existence for a long time, and have been extensively described by various authors (edited by Wexler, 1965). Advances in technology have improved performance, but these descriptions detail the basic techniques.

Materials and procedures

Many factors can influence the best procedures for measuring or controlling water vapour – the range of vapour pressure of interest, the accuracy and time response

Table 4.1 Water contents expressed as vapour pressure, e, saturation deficit, d, dew-point temperature, T_D, and relative humidity, h, at two air temperatures, T_A, and the variation in T_D, h and d associated with a variation of ±0.05 kPa in e, and ±0.5 K in T_A.

					For ±0.05 kPa		For ±0.5 K	
T_A °C	e kPa	d kPa	T_D °C	h %	ΔT_D K	Δh %	Δh %	Δd Pa
20	2.0	0.347	17.4	85.2	±0.4	±2.1	±2.6	±73
20	1.0	1.347	6.9	42.6	±0.7	±2.1	±1.3	±73
20	0.5	1.847	−2.8	21.3	+1.3 −1.4	±2.1	±0.7	±73
20	0.2	2.147	−14.5	8.5	+2.7 −3.4	±2.1	±0.3	±73
5	0.5	0.376	−2.8	57.1	+1.3 −1.4	±5.7	±2.0	±31
5	0.2	0.676	−14.5	22.8	+2.7 −3.4	±5.7	±0.8	±31

required and, often, the relative cost of different systems – but some general principles are appropriate (see also Verdin, 1973).

The materials used for sensor housings, measurement chambers and interconnecting piping should absorb or adsorb as little water as possible. All materials will adsorb some water (see Parkinson, chapter 10, for a review), the minimum quantity measured being about 0.01 μg mm^{-2} on glass (Dixon and Grace, 1982). In general, glass is as good or better than most metals, whose surface finish can influence adsorption; of the plastics, TPX (polymethyl pentene), PTFE and high density polyethylene are in the same class as the metals (Dixon and Grace, 1982; Shepherd, 1973). Acrylic plastic (e.g. "perspex"), which has often been used for chambers in which leaf transpiration is measured, is one of the most unsuitable materials. The significance of water adsorption will be greater in systems with a large surface area of adsorbing material, and where rapid changes in water vapour content are to be measured. This applies particularly in measurements of water potential/activity, where the ratio of chamber surface area to volume is often large. The time taken to reach equilibrium humidity will be greater with more adsorbent materials (Millar, 1971), and will be influenced by chamber geometry and sample size (Dixon and Grace, 1982).

When measuring water vapour in large air volumes or gas streams sensors have often been placed where humidity is to be determined e.g. at particular heights in crop canopies. However if gas sampling is used, pipe runs to the measuring instrument should be as short as possible and care should be taken to avoid condensation in them e.g. by heating the pipes (Bloom *et al.*, 1980). High flow rates through sampling lines will minimise the reponse time. Maintaining the gas flow at slight positive pressure will ensure that any slight leaks do not introduce air – this is particularly necessary with complicated piping systems.

More extensive use of a high accuracy but expensive instrument can be obtained by "time sharing" i.e. switching different gas samples to the instrument in turn. The applicability of this approach depends on the speed of response of the instrument in question and the rate of change of humidity in the various sample lines. Essentially the minimum stabilisation time for the instrument must be much less than the time for significant changes in humidity in any line if direct comparison of each line's humidity is of interest.

Control of humidity

In large volumes or air flows e.g. controlled-environment chambers and wind tunnels, humidity can be adequately controlled using cooling coils to condense water and evaporation pans or steam injectors to increase the water content of the air. These elements are incorporated in a control system which relies upon a measurement of humidity using one of the instruments described in a later section. The accuracy of control is dependent upon the system and the instrument accuracy.

There are a number of techniques available to achieve constant, known humidities in gas streams or confined volumes. The commonest method used is the saturation method in which gas is bubbled through pure water. It is difficult to maintain air at saturation and constant temperature throughout the system is essential. Errors, due either to non-saturation or the carry over of droplets and their subsequent evaporation, can be minimised by maximising the surface area of water, using a train of bubblers, filtering droplets from the air, and limiting the gas flow rate. A two stage method in which the initial saturation is followed by condensation at a lower temperature is better (International Organisation for Standardisation, 1979).

The equilibrium humidities above salt solutions of different concentrations are well documented (Brown and van Haveren, 1972) and are particularly useful with still air e.g. the calibration of instruments to measure water potential. For flowing gases, changes in salt concentration can be avoided by using saturated solutions. Equilibrium humidities for many saturated salt solutions have been measured (Table 4.2). For best results, the salt and water must be pure; the solution must contain plenty of undissolved salt, but crystals must not protrude above the solution's surface; the surface area of solution exposed to the gas should be large. A uniform temperature throughout the system is essential, for though equilibrium relative humidity is only slightly temperature dependent (Table 4.2), the salt solubility varies greatly with temperature. Temperature fluctuations can therefore lead to unsaturation of the solution, especially if it is not stirred.

Just as an equilibrium humidity is established when saturated solution and salt crystals are both present, a similar equilibrium exists between air and pairs of salt hydrates (Richardson and Malthus, 1955). The advantage over saturated solutions is that by using solid chemicals it provides a cleaner, more portable system (Parkinson and Day, 1981). The equilibrium humidities for a number of hydrate pairs are given in Table 4.3.

A range of lower humidities can be obtained by combining any of these generation methods with a gas mixing system. Known flow rates of dry and humidified air can be combined either by using calibrated flow meters and needle valves, or critical flow

Table 4.2 Equlibrium relative humidities (%) over saturated salt solutions. Data are selected from Greenspan's (1977) review of the literature. The uncertainties are estimated by Greenspan from many experimental data.

	10°C	20°C	30°C
CsF	4.9±1.6	3.8±1.1	3.0±0.8
LiBr	7.1±0.7	6.6±0.6	6.2±0.5
ZnBr	8.5±0.7	7.9±0.5	7.6±0.3
KOH	12.3±1.4	9.3±0.9	7.4±0.6
NaOH	–	8.9±2.4	7.6±1.7
LiCl	11.3±0.4	11.3±0.3	11.3±0.2
$CaBr_2$	21.6±0.5	18.5±0.5	–
LiI	20.6±0.3	18.6±0.2	16.6±0.1
$K(CH_3COO)$	23.4±0.5	23.1±0.3	21.6±0.5
$MgCl_2$	33.5±0.2	33.1±0.2	32.4±0.1
NaI	41.8±0.8	39.7±0.6	36.2±0.4
K_2CO_3	43.1±0.4	43.2±0.3	43.2±0.5
$Mg(NO_3)_2$	57.4±0.3	54.4±0.2	51.4±0.2
NaBr	62.2±0.6	59.1±0.4	56.0±0.3
$CoCL_2$	–	–	61.8±2.8
KI	72.1±0.3	69.9±0.3	67.9±0.2
$SrCl_2$	75.7±0.1	72.5±0.05	69.1±0.03
$NaNO_3$	77.5±0.5	75.4±0.4	73.1±0.3
NaCl	80.6±1.0	79.2±0.4	77.9±0.6
KBr	83.8±0.2	81.7±0.2	80.3±0.2
$(NH_4)_2SO_4$	82.1±0.5	81.3±0.3	80.6±0.3
KCl	86.8±0.4	85.1±0.3	83.6±0.3
$Sr(NO_3)_2$	90.6±0.4	86.9±0.3	–
KNO_3	96.0±1.4	94.6±0.7	92.3±0.6
K_2SO_4	98.2±0.8	97.6±0.5	97.0±0.4
K_2CrO_4	–	–	97.1±0.4

orifices (Parkinson and Day, 1981). A range of humidities can also be obtained by using the two pressure standard (Amdur and White, 1965). Air is saturated at total pressure p_1 and then depressurised to pressure p_2. Assuming ideal gas behaviour, the resulting relative humidity

$$h = p_2/p_1 \ [\text{or} \ 100 p_2/p_1 \%] \qquad 4.4$$

The error due to non-ideal behaviour is less than 0.3%. Hasegawa *et al.* (1965) have considered the errors involved in this method of humidity generation.

The removal of all water from an air stream is also a common requirement. Table 4.4 summarises some often quoted data to indicate the suitability and drying power of various chemical desiccants. Physical drying using cooling coils or solid CO_2 (–73°C) or liquid nitrogen (–196°C) to condense water vapour may also be used. When CO_2 concentration of the air stream is being measured or controlled, the drying agent used

Table 4.3 Equilibrium relative humidities (%) over pairs of salt crystal hydrates; data from Wilson (1920), Baxter and Lansing (1920), Schumb (1923), Baxter and Cooper (1924), Montgomery *et al.* (1965), Thakker *et al.* (1968) and Parkinson and Day (1981).

	15°C	23°C	25°C	30°C
$Mg(ClO_4)_2.2H_2O/4H_2O$		0.039		
$Mg(ClO_4)_2.4H_2O/6H_2O$		0.099		
$LiCl/LiCl.H_2O$		2.0		
$LiNO_3/LiNO_3.\frac{1}{2}H_2O$				2.6
$CuSO_4/CuSO_4.H_2O$			3.4	
$H_2C_2O_4/H_2C_2O_4.2H_2O$	9.0		11.2	
$LiOH/LiOH.H_2O$				16.9
$CuSO_4.H_2O/3H_2O$			23.6	
$BaCl_2.H_2O/2H_2O$	22.1		23.9	
$Na(CH_3COO)/Na(CH_3COO).3H_2O$	23.6		26.9	
$CuSO_4.3H_2O/5H_2O$			32.8	
$SrCl_2.2H_2O/6H_2O$	32.2		35.5	
$MgSO_4.6H_2O/7H_2O$			52.6	
$FeSO_4.4H_2O/7H_2O$	52.6		60.2	
$Na_2HPO_4.2H_2O/7H_2O$			61.1	
$ZnSO_4.6H_2O/7H_2O$			64.6	
$CoSO_4.6H_2O/7H_2O$			69.8	
$Na_2CO_3.7H_2O/10H_2O$	68.1		75.8	
$Na_2HPO_4.7H_2O/12H_2O$	69.9		80.3	
$Na_2SO_4/Na_2SO_4.10H_2O$	72.0		80.8	
$NiSO_4.6H_2O/7H_2O$			87.1	

should absorb as little CO_2 as possible. Silica gel is particularly unsuitable, "Drierite" ($CaSO_4$) may be used though it requires some time to equilibrate (Koller and Samish, 1964), anhydrone ($Mg(ClO_4)_2$) is better, but where possible drying by cooling is to be preferred (Samish, 1978).

Measurement of humidity

The instruments available are listed in Table 4.5, grouped according to their absolute accuracy and the closeness of the relation between the measurement principle and the physical properties of water vapour (the lower the group number the closer the relation and the greater the accuracy).

The various methods of humidity generation are included as they may be used to calibrate instruments in higher numbered groups. The absolute standard for humidity measurement is the gravimetric hygrometer (Wexler and Hyland, 1965) which involves absorption of water vapour by a solid desiccant, the precise weighing of the increase in mass, and the precise determination of the volume of the associated gas of known density. It is not suitable as a working laboratory or field instrument, but is

Table 4.4 Drying capabilities of some chemical desiccants (data from Luck (1964), Trussel and Diehl (1963), quoted by Slavik (1974), and a desiccant selector chart produced by J.T. Baker Chemicals, Deventer, Holland). The residual values will depend on flow rate and chemical purity.

	Residual water $g\ m^{-3}$	Dew-point °C	Water capacity of desiccant $g\ g^{-1}$	Regeneration
Anhydrous $Mg(ClO_4)_2$	0.0002	−87		
'Anhydrone' ($Mg(ClO_4)_2$)	0.001	−78	0.2	250°C in vacuum
P_2O_5	0.001–0.004	−78 to −69	0.5	No
Molecular sieve (various)	0.001–0.003	−78 to −71	0.18	250°C
Al_2O_3	0.003	−71	0.2	175°C
H_3PO_4	0.003	−71		No
H_2SO_4	0.003–0.004	−71 to −69		No
'Drierite' ($CaSO_4$)	0.005	−67	0.07	200–225°C
CaO	0.007–0.2	−65 to −36	0.3	1000°C
MgO	0.008	−64	0.5	800°C
Silica gel	0.03	−52	0.2	200–350°C
Anhydrous $CaCl_2$	0.07	−46	0.2–0.3	No
$CaCl_2$ (various)	0.14–0.36	−40 to −30	0.2–0.3	No
NaOH	0.16	−39		No
KOH	0.3	−32		No
$ZnCl_2$	0.9	−20		110°C
$MgSO_4$	1.0	−19		No
$CuSO_4$	1.4	−15		

capable of high accuracy – 1 part in 10^3 for specific humidities between 0.2 and 30 mg g^{-1}. Most of the other instruments have been or could be used in environmental physiology.

Condensation dew-point hygrometers. These instruments can operate over the whole range of interest for environmental humidity measurement, and their calibration stability makes them the most suitable laboratory calibration standard. In operation (Fig. 4.1) a mirror exposed to the sample gas stream is cooled until changes in the light reflected from the mirror surface indicate dew formation. The principle sources of error are temperature gradients between the water surface and the thermometer, effects of droplet size and the uncertain condition of the condensed layer, and temperature measurement. If the mirror is contaminated with soluble salts then the temperature determined will be higher than the true dew-point temperature of the air. At temperatures below 0°C, a frost layer will be stable, but initially a metastable dew layer will be formed, and this can be long-lived even down to temperatures of −40°C. For accurate determinations in this temperature range it is necessary therefore to be able to view the mirror through a microscope. The accuracy achievable is about ±0.2 K dew-point, but to be sure of this accuracy, periodic checks against another

Table 4.5 Humidity measurement or generation instruments in groups that reflect the accuracy and the nature of the operation involved.

Group I		Gravimetric hygrometer		
Group II	Condensation dew-point hygrometer	Two pressure generator		Saturation generator
Group III	Pneumatic bridge hygrometer	Wet bulb psychrometer	Saturated salt solutions (generator)	Salt hydrate pairs (generator)
	Spectroscopic (UV or IR) hygrometers	Electrolytic hygrometer (P_2O_5)		Mechanical (hair) hygrometer
Group IV	Hygroscopic sensors (resistance/ capacitance)		Piezoelectric hygrometer	

calibration standard are advisable. For fast response to humidity changes, the instrument should have a small internal volume and no dead airspaces. It is also often useful to have a remote sensor unit that can be placed close to the sampling point.

Most dew-point hygrometers are engineered for the measurement of water vapour in flowing gas streams. One recent addition is designed for measuring equilibrium humidity in still air, particularly for the measurement of water potential in solids such as seeds and food products. The heat dissipation associated with cooling the mirror is minimised by cycling the mirror temperature, rather than maintaining its temperature at the dew-point. This also reduces the risk of contaminants accumulating on the mirror (Easty and Young, 1976). In another novel instrument, Regtien and Makkink

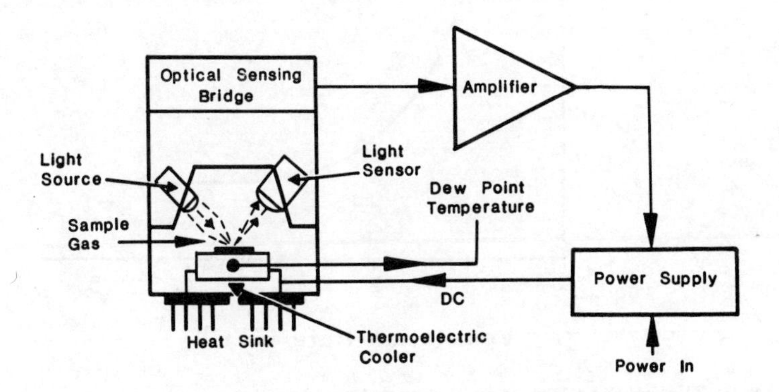

Figure 4.1 The configuration of the measurement system in a dew-point hygrometer (redrawn from EG & G sales literature).

(1978) have eliminated the need for an optical system to detect the dew formation by sensing the capacitance changes in an evaporated film capacitor when dew forms. The films are evaporated on a silicon chip whose surface temperature is determined by measuring the base-emitter voltage of a transistor located in the centre of the chip. This approach may be useful in dusty or polluted environments.

Wet and dry bulb psychrometers. These are probably still the most frequently used instruments for humidity measurement. The principle of operation is that evaporation of water from a wick surrounding one thermometer requires energy and the thermometer cools relative to a dry thermometer. The vapour pressure can be calculated from the temperature depression, $(T_A - T_W)$, using

$$e = e_s(T_W) - Ap(T_A - T_W) \qquad 4.5$$

where A is a constant and p the total atmospheric pressure. Monteith (1954) has assessed some of the sources of error in practical use of this equation. At high ventilation rates, Ap is equal to the psychrometric constant $\gamma = c_p p / \lambda \epsilon$, where c_p is the specific heat of air at constant pressure, λ is the latent heat of vaporisation of water and ϵ is the ratio of molecular weights of water and air (0.622). At atmospheric pressure γ is 66 Pa K^{-1}, but this value only strictly applies when the heat exchange resistance of the wet bulb is equal to its vapour exchange resistance (Monteith, 1973). Monteith gives an example for an Assman psychrometer with a 3 mm wet bulb diameter and 3 m s^{-1} ventilation rate, for which the true Ap value is 63 Pa K^{-1}, tending to 59 Pa K^{-1} at higher ventilation rates. At lower ventilation rates, the increasing importance of radiative heat exchange in determining the wet bulb temperature (decreasing the wet bulb depression) leads to an increasing psychrometric constant

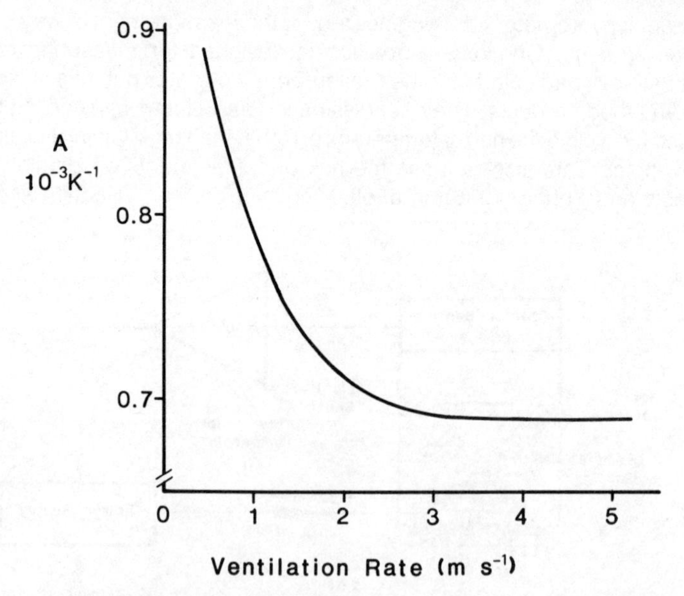

Figure 4.2 The relation between the constant A in the psychrometer equation and ventilation rate for an 8 mm diameter wet bulb (redrawn from Meteorological Office (1981)).

(Fig. 4.2). Heat conduction along the stem of the wet bulb thermometer to its cooled sensing point will also tend to decrease the wet bulb depression. For accurate use of such psychrometers it is best to ensure that Ap is as close to the psychrometric constant as possible: heat conduction to the thermometers should be minimised e.g. by using fine wires to thermocouple thermometers, or a narrow cross section glass thermometer; small external diameter thermometers should be used also to minimise radiative heat exchange; and a high ventilation rate should be used. For accurate work, the deviation of Ap from γ should be determined either theoretically (Monteith, 1973) or by calibration against a standard instrument. Other sources of error are dirt on the wick or in the water supply, a poorly fitting sleeve or one with inadequate wetting properties, and failure to shield the thermometer from direct short-wave radiation. The errors associated with temperature measurements are considered by Bell and Rose (Chapter 5). In careful use, the absolute accuracy that can be achieved with a wet and dry bulb psychrometer will be $\pm 2\%$ saturation deficit or ± 0.15 K wet bulb depression, whichever is the greater.

Spectroscopic methods. Non-dispersive infra-red gas analysers (see Chapter 3) have been used for measuring water vapour for many years (Slavik, 1974). The radiation used may be broad-band, or a particular frequency may be selected using narrow-band interference filters. After passage through the analyser tube that contains the sample gas, the transmitted radiation is detected either using a selective detector such as the Luft detector filled with NH_3 gas, which has a similar absorption spectrum to that of water vapour (Slavik, 1974), or using a fast responding pyroelectric detector if the radiation is filtered. Among recent instruments commercially available are open path analysers for *in situ* measurements, and portable analysers (see Chapter 3 for a description of these types of analyser).

Radiation absorption in the ultra violet range can also be used for water vapour measurements. The most sophisticated techniques can distinguish between absorption by water and by O_2, by illuminating at 121.6 nm (the Lyman α line) and measuring the resultant fluorescence at 310 nm from activated water molecules (Kley and Stone, 1978; Stone, 1980). The transmission, τ, of 121.6 nm radiation through a cell of length l mm is approximately

$$\tau = \exp(-l\,(2.5\,10^{-2} p[O_2] + 25.0 p[H_2O])) \qquad 4.6$$

where $p[O_2]$ and $p[H_2O]$ are the partial pressure in kPa (Kley and Stone, 1978). In air at atmospheric pressure, the absorption by O_2 is equal to that of 0.02 kPa water vapour, so Lyman α absorption measurement is sufficient for the physiologist's range of interest.

Both types of instrument are quite expensive, and need direct calibration. At the vapour pressures of interest there is considerable self-broadening of the absorption bands which is proportional to concentration (Parkinson and Legg, 1971); transmission is not linearly related to concentration particularly for the ultra violet instrument (with a 2 mm cell length, Lyman α transmission for 1 kPa vapour pressure is 0.007). Accuracy is therefore dependent on good calibration. The differential mode for the IRGA (Parkinson and Legg, 1971) is particularly useful for the resolution of small differences against a large background (say 10 Pa in 2 kPa).

Hygroscopic (resistance/capacitance change) sensors. There are many commercial instruments available based on a wide variety of sensors of this general type. They range from being slow responding and of dubious dependability to fast responding sensors of good accuracy. Many of the sensors referred to by Wexler (1965) and Slavik (1974) are still available. Some of the recent developments amongst these sensors have found extensive use in environmental physiology.

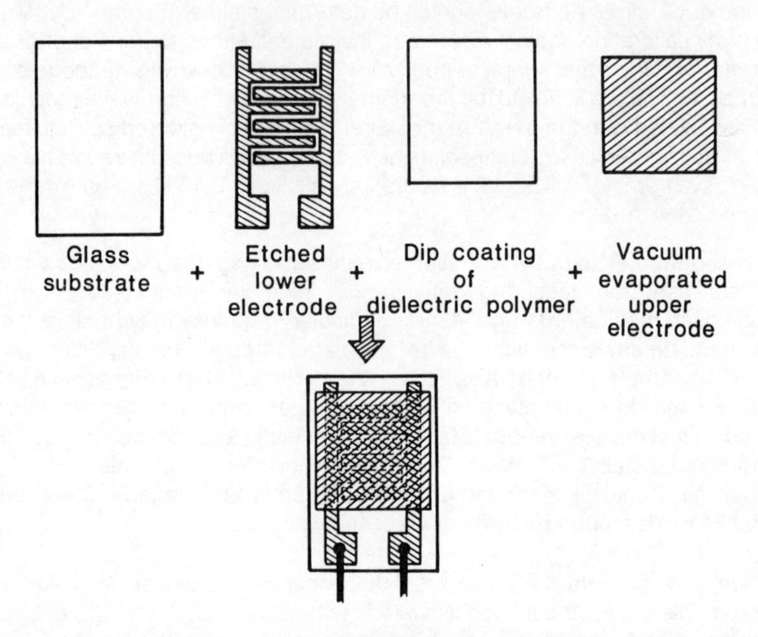

Figure 4.3 The construction of the dielectric polymer humidity sensor (Vaisala "Humicap" – redrawn from Salasmaa and Kostamo (1982)).

The Vaisala dielectric polymer sensor became available in 1973 and has found many applications: it consists of a polymer layer coated onto a metallised glass plate, with a second gold electrode vacuum-evaporated onto the polymer surface (Fig. 4.3). The upper electrode is thin so that water can diffuse through it rapidly. The electrical capacitance of the dielectric changes almost linearly with relative humidities from 0-75%. The major attractions of the sensor are small size (4 x 6 x 1 mm, approximately), linearity of response, small temperature dependence, and rapid response to changes in humidity (Salasmaa and Kostamo, 1982). This response has a fast and slow phase, both of which are slowed as the temperature falls (Fig. 4.4) but 90% response to a humidity change is achieved in about 1 s at 20°C. The response is less satisfactory at high relative humidities, above about 75%, when long term drift over many hours occurs. When used below this humidity, an absolute accuracy of ±2% relative humidity is readily achieved. One way to maintain this accuracy to higher humidities is to heat the sensor so that it never experiences relative humidities in excess of 75% – this technique is commercially available.

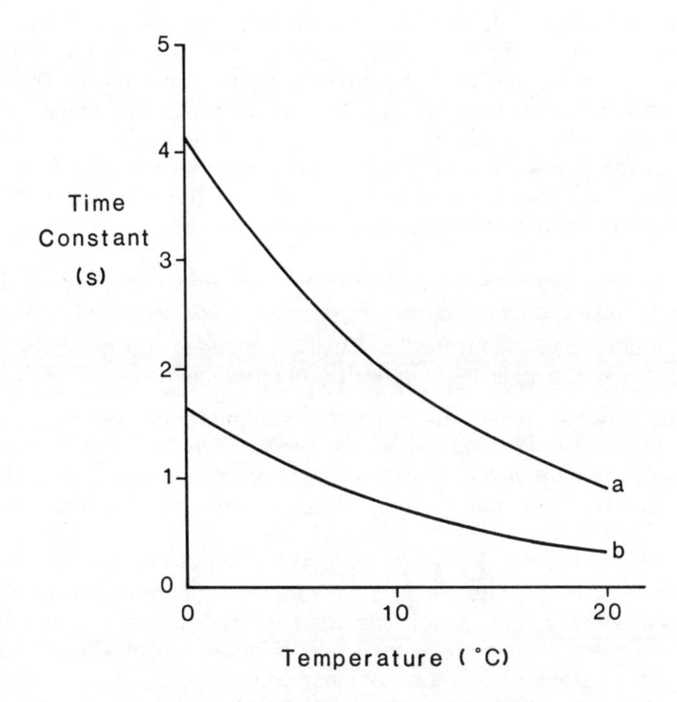

Figure 4.4 The temperature dependence of the response time of a Vaisala 'Humicap' sensor for (a) 90% or (b) 60% response to an imposed humidity change (redrawn from Salasmaa and Kostamo (1982)).

The technology of dielectric polymer sensors is likely to continue to change. Certainly a number of sensors are being produced now with similar characteristics – the variations in design are small but often significant. One has a thick porous external electrode layer of chrome rather than a thin gold layer. It is claimed that the thick layer provides greater protection to the polymer, but allows easy penetration of water vapour via the pores. Others claim stability up to 100% relative humidity. Of particular interest is a product from Mullard that uses a non-conducting foil coated with gold on each side as the capacitance detector. It can be used over the full humidity range with an accuracy of about ±3% relative humidity, and though not fast responding (90% response to humidity change in about 4 minutes) it is far cheaper than most other sensors (S. W. Burrage, pers. comm.).

There are many other instruments which fall into the category of hygroscopic sensors. Sulphonated polystyrene and lithium chloride are commonly used in sensors whose resistance-change with humidity is measured. Lithium chloride is also used in the "dewcel" type sensor; the action is based on the great affinity of the salt for water and the change in resistance that occurs when the ambient humidity changes from below to above the equilibrium humidity of the saturated salt solution (Table 4.1). A voltage applied across the sensor causes its temperature to rise: the rate of heating falls off markedly when the sensor temperature reaches the point at which the relative humidity experienced by the salt is less than the equilibrium humidity. This technique is only usable at relative humidities above the equilibrium humidity for saturated lithium chloride (Table 4.1). Aluminium oxide is commonly used in sensors where capacitance-change is measured. The aluminium oxide layer in such sensors is very thin, often only a few microns thick, so equilibrium with the air is established rapidly, and sensor response time is short (typically less than 10 s).

The response characteristics of all of these sensors are too varied to give in detail. Some are quite fast responding, others show considerable hysteresis, and all are likely to require regular calibration checks. Their advantages are generally related to robustness, cheapness or ability to work in polluting atmospheres.

Other instruments. There are many techniques that can be used to measure water vapour. Some are particularly appropriate to the measurement of low humidity outside the range of interest of most environmental physiologists, some have been superseded, and some are not generally available in an operational instrument.

The pneumatic bridge hygrometer (Greenspan, 1965; Wildhack *et al.*, 1965) is not available commercially. Its operation is based on the change in volume flow rate of a moist gas when its water vapour is absorbed by a chemical desiccant. The change in flow rate is detected using a "Wheatstone" bridge arrangement with critical-flow orifices as the resistance elements – for these orifices the flow rate is proportional to the absolute pressure on the input side. A theoretical relation between out-of-balance pressure difference and water vapour content can be determined based on the orifice characteristics. Accuracies of 0.3% relative humidities are claimed.

Electrolytic hygrometers measure the current that flows in the electrolysis of water absorbed on a phosphorous pentoxide sensor. If the sensor absorbs all the water in the air, the current required for electrolysis is theoretically defined by Faraday's law (Keidel, 1959) and measurement accuracy is largely limited by the accuracy with which the gas flow rate is measured. However, complete absorption is only achieved at low vapour pressures and flow rates (typical maximum about 0.1 kPa at 1.6 $cm^3 s^{-1}$), and though diffusers or low flow rate sample lines extend the concentration range, these instruments are principally intended for very low humidity applications.

The piezoelectric hygrometer measures changes in the oscillation frequency of a quartz crystal oscillator when water is absorbed by a hygroscopic coating on the crystal (King, 1965). Quartz crystals are used as microbalances in many applications e.g. vacuum evaporation, and have a high sensitivity, the addition of 1 ng mm^{-2} being detectable. The sophisticated commercial instruments available are capable of detecting 2 10^{-3} Pa, and can be used throughout the humidity range. However, their accuracy is only 5-10% being crucially dependent on the calibration, as the frequency shift is a non-linear function of vapour concentration. The minimum response time, when humidity is increasing, is limited by the mode of operation which involves switching sample and dry gas streams alternately over the crystal for 30 s periods.

Finally the mechanical hygrometer, based on changes in length of hair with humidity, must be referred to as it is still the commonest source of general humidity measurement (Davey, 1965). The human hair has been used for centuries, but though it has advantages in cheapness and simplicity, it is not accurate or reproducible. The length change is approximately logarithmically related to relative humidity, but shows considerable hysteresis (about 3% relative humidity) and temperature response. The response time is 5 min or more, and the calibration can be permanently shifted if the hair is exposed to relative humidities less than 15% or greater than 90%. Ageing and biodeterioration also modify the calibration, so the hair hygrometer is only suitable for approximate measurements.

Practical considerations and applications

When water vapour is to be measured in physiological experiments, there are many considerations other than the absolute accuracy of observations that will determine the optimum instrument and measurement configuration. For flexibility and high accuracy, the condensation dew-point hygrometer is probably of most general value, though it is expensive. The dielectric sensors are small, require quite simple electronics and therefore are particularly suitable for use in portable instruments (Day, 1977) or where multi-point measurement is required. The thermocouple psychrometer has been extensively developed for the measurement of equilibrium relative humidity to determine water potential/activity, and the sample chamber versions of these instruments seem particularly well designed (Wiebe, 1981).

When experiments require data for long periods or for many sites, some less than perfect methods are often used. Hair hygrographs are commonly used to provide continuous records of humidity, and so reference to them may be made when generalisations are required from more detailed studies that have employed more suitable techniques e.g. comparisons between microclimate and local weather (Teel *et al.*, 1982). Neither is it surprisimg that they find continued use in other biological studies e.g. entomology (Boyne and Hain, 1983) where more suitable instrumentation is not available. However, it is disconcerting that in environmental plant physiology, where quantitative studies of water exchange rates and plant stress are being made, these instruments are still being used (O'Toole and Baldia, 1982).

Wet and dry bulb psychrometers are used in many studies of micro- and meso-climate. The simplicity of their operation, relying solely on temperature measurements, is a great attraction, but there are many sources of error that should be noted (see above). Forced ventilation of the wet bulb is the only way to ensure a well defined psychrometric constant. When power for a fan or pump is not available for ventilation, the value used for the psychrometric constant may be a major source of error. Standard meteorological observations of humidity rely upon measurements of the depression in temperature of a wet bulb mercury-in-glass thermometer placed inside a Stevenson screen. Not only is there no forced ventilation, but the screen substantially impedes natural ventilation. The value of A used in standard meteorology is 0.799mK^{-1}–reflecting the inadequate ventilation–but the true value on any particular occasion may deviate from this by 10% or more. At the lowest ventilation rates, a more stable psychrometric constant may apply but this will be at the expense of speed of response to changing humidity. These considerations of ventilation are particularly appropriate when field measurements are made of the profile of humidity above the ground surface, as windspeed and hence ventilation will vary with psychrometer position. Ventilation will be particularly small if the lowest psychrometers are within the crop canopy, but in these circumstances high ventilation rates may not be appropriate if they disturb the natural conditions that are being measured. It may then be best to determine actual windspeed and wet bulb depression, and use a theoretical or empirical relation between the psychrometric constant and windspeed for accurate humidity determination.

The Assman clockwork aspirated wet and dry bulb psychrometer is widely used as a transfer standard in the calibration of other humidity instruments or checking of controlled environments. If care is taken to ensure optimum operation and exposure, such instruments provide a simple portable system for calibrations, but they will not have as good an absolute accuracy as the condensation dew-point meter, particularly at low humidities.

It should be clear from these various practical considerations that there are many different approaches that may be taken to the measurement of humidity and that each has drawbacks as well as advantages. Of all the physical variables that the environmental physiologist may wish to determine, humidity poses the most problems and is measured with the least accuracy. Many of the important factors in making accurate measurements have been described above – but the need for regular and careful calibration deserves further stressing. All instruments are prone to some calibration drift, particularly those used in dusty or polluting environments, and checks against a local transfer standard, e.g. a dew-point hygrometer, should be part of the measurement routine: even that transfer standard will need to be checked occasionally. The National Bureau of Standards (NBS) in the U.S.A have sophisticated facilities for calibration, and many dew-point hygrometers are sold with calibration certificates traceable to the NBS. In the UK a humidity standard is currently being established at the National Physical Laboratory (J. L. Hales, pers. comm.). It is based on two devices, a gravimetric hygrometer and a two temperature saturation generator, and aims to give 0.5% accuracy for relative humidities in the range 10-98% and 0.05 K dew-point accuracy, for dew-point temperatures from –70°C to +90°C and at a range of air pressures.

References

Amdur, E.J. & White, R.W. (1965). Two-pressure relative humidity standards. *In* Humidity and Moisture. Vol. III. ed. A. Wexler, pp.445-454. New York: Reinhold Publishing Corp.

Baxter, G.P. & Cooper, W.C. (1924). The aqueous pressure of hydrated crystals. II. Oxalic acid, sodium sulfate, sodium acetate, sodium carbonate, disodium phosphate, barium chloride. J. Am. chem. Soc., **46**, 923-933.

Baxter, G.P. & Lansing, J.E. (1920). The aqueous pressure of some hydrated crystals. Oxalic acid, strontium chloride, sodium sulfate. J. Am. Chem. Soc., **42**, 419-426.

Bloom, A.J., Mooney, H.A., Bjorkman, O. & Berry, J.A. (1980). Materials and methods for carbon dioxide and water exchange analysis. Plant, Cell & Environ., **3**, 371-376.

Boyne, J.V. & Hain, F.P. (1983). Effects of constant temperature, relative humidity and simulated rainfall on development and survival of the spruce spider mite (*Oligonychus ununguis*). Can. Ent., **115**, 93-105.

Brown, R.W. & van Haveren, B.P. (1972). Psychrometry in water relations research. Proceedings of the Symposium on thermocouple psychrometers. Utah State University, March 1971. pp.342 . Utah Agricultural Experiment Station, XII.

Buck, A.L. (1981). New equations for computing vapour pressure and enhancement factor. J. appl. Meteorol., **20**, 1527-1532.

Davey, F.K. (1965). Hair humidity elements. *In* Humidity and Moisture. Vol. I. ed. A. Wexler, pp.571-573. New York: Reinhold Publishing Corp.

Day, W. (1977). A direct reading continuous flow porometer. Agric. Meteorol., **18**, 81-89.

Day, W., Lawlor, D.W. & Legg, B.J. (1981). The effects of drought on barley: soil and plant water relations. J. agric. Sci., Camb., **96**, 61-77.

Dixon, M. & Grace, J. (1982). Water uptake by some chamber materials. Plant, Cell & Environ., **5**, 323-327.

Easty, A.C. & Young, S. (1976). A small scale dewpoint humidity measurer. J. Phys., E. (Scientific Instruments), **9**, 106-110.

Greenspan, L. (1965). A pneumatic bridge hygrometer for use as a working humidity standard. *In* Humidity and Moisture. Vol. II. ed. A. Wexler, pp.433-443. New York: Reinhold Publishing Corp.

Greenspan, L. (1977). Humidity fixed points of binary saturated aqueous solutions. J. Res. natn. Bur. Stand., **81**A, 89-96.

Hasegawa, S., Hyland, R.W. & Rhodes, S.W. (1965). A comparison between The National Bureau of Standards two pressure humidity generator and the National Bureau of Standards Standard Hygrometer. *In* Humidity and Moisture, Vol. III ed. A. Wexler, pp.455-459. New York: Reinhold Publishing Corp.

International Organisation for Standardisation (1979). Gas analysis – preparation of calibration gas mixtures – saturation method. ISO 6147 - 1979 (E) 2 pp.

Keidel, F.A. (1959). Determination of water by direct ampherometric measurement. Analyt. Chem., **31**, 2043-2046.

King, W.H. (1965). The piezoelectric sorption hygrometer. *In* Humidity and Moisture, Vol. I. ed. A. Wexler, pp. 578-583. New York: Reinhold Publishing Corp.

Kley, D. & Stone, E.J. (1978). Measurement of water vapour in the stratosphere by photodissociation with Ly α(1216Å) light. Rev. scient. Instrum., **49**, 691-697.

Koller, D. & Samish, Y.B. (1964). A null point compensating system for simultaneous and continuous measurements of net photosynthesis and transpiration by controlled gas-stream analysis. Bot. Gaz., **125**, 81-88.

Luck, W. (1964). Feuchtigkeit. Munchen-Wien: Oldenburg.

Meteorological Office (1981). Handbook of Meteorological Instruments Volume 3. Measurement of Humidity. London: HMSO.

Millar, B.D. (1971). Improved thermocouple psychrometer for the measurement of plant and soil water potential. I. Thermocouple psychrometry and an improved instrument design. J. exp. Bot., **22**, 875-890.

Monteith, J.L. (1954). Error and accuracy in thermocouple psychrometry. Proc. phys. Soc. Lond., Ser. B, **67**, 217-226.

Monteith, J.L. (1973). Principles of Environmental Physics. London: Arnold.

Montgomery, C.D., Googin, J.M. & Phillips, L.R. (1965). Moisture monitor testing and calibration. *In* Humidity and Moisture, Vol. III. ed. A. Wexler, pp.467-471. New York: Reinhold Publishing Corp.

O'Toole, J.A. & Baldia, E.P. (1982). Water deficits and mineral uptake in rice. Crop Sci., **22**, 1144-1150.

Parkinson, K.J. (1971). Carbon dioxide infra-red gas analysis. Effects of water vapour. J. exp. Bot., **22**, 169-176.

Parkinson, K.J. & Day, W. (1981). Water vapour calibration using salt hydrate transitions. J. exp. Bot., **32**, 411-418.

Parkinson, K.J. & Legg, B.J. (1971). A new method for calibrating infra-red gas analysers. J. Phys., E. (Scientific Instruments), **4**, 589-600.

Rawson, H.M., Begg, J.E. & Woodward, R.G. (1977). The effect of atmospheric humidity on photosynthesis, transpiration and water use efficiency of leaves of several plant species. Planta, **134**, 5-10.

Regtien, P.P.L. & Makkink, H.K. (1978). A capacitive dew-point sensor. Delft Prog. Rep., B., **3**, 107-110.

Richardson, G.M. & Malthus, R.S. (1955). Salts for static control of humidity at relatively low levels. J. appl. Chem., Lond., **5**, 557-567.

Salasmaa, E. & Kostamo, P. (1982). “Humicap” thin film humidity sensor. Vaisala, 1982-05-10.

Samish, Y.B. (1978). Measurement and control of CO_2 concentration of air is influenced by the desiccant. Photosynthetica, **12**, 73-75.

Schumb, W.C. (1923). The dissociation pressure of certain salt hydrates by the gas-current saturation method. J. Am. chem. Soc., **45**, 342-354.

Shepherd, W. (1973). Moisture absorption by some instrument materials. Rev. scient. Instrum., **44**, 234.

Slavik, B. (1974). Methods of studying plant water relations. Berlin: Springer-Verlag.

Smithsonian Institution (1968). Smithsonian meteorological tables. Sixth edition (R.J. List). Washington: Smithsonian Institution Press.

Stone, R.J. (1980). Ultraviolet fluorescence water vapour instrument for aircraft. Rev. scient. Instrum., **51**, 677-678.

Teel, P.D., Fleetwood, S.C. & Heubner, G.L. (1982). An integrated sensing system designed for unattended continuous monitoring of micro-climate relative humidities and its use to determine the influence of vapour pressure deficits on tick (*Acari ixodoidea*) activity. Agric. Meteorol., **27**, 145-154

Thakker, M.T., Chi, C.E., Peck, R.E. & Wasan, D.T. (1968). Vapour pressure measurements of hygroscopic salts. J. chem. Engng. Data, **13**, 553-558.

Trussel, F. & Diehl, H. (1963). Efficiency of chemical desiccants. Analyt. Chem., **35**, 674-677.

Verdin, A. (1973). Gas analysis instrumentation. London: MacMillan.

Wexler, A. (1965). Humidity and Moisture, Vol. I., II. & III. New York: Reinhold Publishing Corp.

Wexler, A. (1976). Vapour pressure formulation for water in the range 0 to 100°C – A Revision. J. Res. natn. Bur. Stand., **80**A, 775-ff.

Wexler, A. & Hyland, R.W. (1965). The NBS standard hygrometer. *In* Humidity and Moisture, Vol. III. ed. A. Wexler, pp. 389-432. New York: Reinhold Publishing Corp.

Wiebe, H.H. (1981). Measuring water potential (activity) from free water to oven dryness. Pl. Physiol., **68**, 1218-1221.

Wildhack, W.A., Perls, T.A., Kissinger, C.W. & Hayes, J.W. (1965). Continuous-absorption hygrometry with a pneumatic bridge utilising critical flow. *In* Humidity and Moisture, Vol. I. ed. A. Wexler, pp.552-570. New York: Reinhold Publishing Corp.

Wilson, R.E. (1920). Some new methods for the determination of the vapour pressure of salt-hydrates. J. Am. chem. Soc., **41**, 704-725.

Chapter 5: The measurement of temperature

C. J. Bell and D.A. Rose
Glasshouse Crops Research Institute, Worthing Road,
Littlehampton, West Sussex, BN17 6LP, England.

Introduction

What is temperature?

In everyday experience temperature is ubiquitous. We feel either hot, cold or neutral as does every object we touch. But what are we sensing and how can we assign a scale to it and measure it?

Temperature is an inherent property of a body. We observe that it is related to heat and hence energy, because if we add heat to a 'cold' body we can produce a 'hot' body. We can also remove heat from a body, but there is no one-to-one correspondence between heat and temperature. The same amount of heat added to different bodies will in general produce different sensations of temperature.

To define temperature we must look beyond our subjective sensations to the underlying principles. Thermal equilibrium is a concept we can understand and define. Two bodies in contact are in thermal equilibrium if neither gains nor loses heat. We can then say that their temperatures are equal. Futhermore, we can postulate that two bodies both in thermal equilibrium with a third body are in thermal equilibrium with each other. This postulate is called *the zeroth law of thermodynamics* and forms the basis of the physical concept of temperature.

Temperature, therefore, is a macroscopic property of a body which determines whether it will be in thermal equilibrium with another body when they are brought into contact. If not, one body will lose heat and the other gain it. How then do we measure temperature?

Temperature scales

Many physical properties vary as our physiological perception of temperature changes. Any of these can be used to measure temperature and each will lead to a practical temperature scale. Each scale can be calibrated against the others but none has any particular claim to be the 'right' scale.

Thermodynamic temperature is a physical quantity based on the concept of the Carnot engine and on statistical mechanics (see, for example, Halliday and Resnick, 1978). It is measured and expressed as a coefficient times a unit. The unit is the kelvin (after Lord Kelvin), symbol K, and defined as the fraction 1/273.16 of the thermodynamic temperature of the triple point of water (the point at which ice, water and water vapour co-exist in equilibrium). All other thermodynamic temperatures are determined by their ratio to this assigned value.

When temperature is used in thermodynamics or statistical mechanics it is the thermodynamic temperature that is implied. The gas thermometer is the most accurate practical instrument to measure this. Measurements using different gases give similar results and lead to the concept of the *ideal gas scale*, usually achieved using helium in a constant-volume gas thermometer. Although it can be shown that the ideal gas temperatures are close to thermodynamic temperatures, the gas thermometer is impractical for common use. In 1927, the Seventh General Conference on Weights and Measures decided to separate the idea of thermodynamic temperature from the need to have a practical temperature scale which could be realised in most laboratories. In its most recent form, this is the International Practical Temperature Scale of 1968 (IPTS-68), amended in 1975 (National Physical Laboratory, 1976). It consists of 11 defining fixed points which have highly reproducible temperatures to which values have been assigned which conform as closely as possible to thermodynamic temperature. These range from the triple point of hydrogen at 13.81 K to the normal freezing point of gold at 1337.58 K and include the triple point of water at 273.16 K and the normal boiling point of water at 373.15 K. The last two are the only defining fixed points relevant to the normal biological range of temperature.

Between the fixed points IPTS-68 specifies the use of certain instruments, mostly platinum resistance thermometers, to interpolate using given formulae. Thus IPTS-68 deviates from thermodynamic temperature between the fixed points, but the differences are negligible for biological studies. There are also 23 secondary fixed points whose temperatures have been determined by the specified interpolation instruments. Of these, the freezing point of mercury at 234.288 K, the ice point at 273.15 K and the triple point of phenoxybenzene at 300.02 K are relevant to biology. (The melting point of gallium at 302.92 K is also considered to be a temperature standard of interest to biologists by the National Bureau of Standards, who supply calibration cells containing this material).

IPTS-68 is also expressed in kelvins, but these vary from the thermodynamic kelvins depending where on the scale we are working. However, the maximum deviation is only 0.4% and is thus negligible in biology. An alternative unit is also allowed, the degree Celsius (named after the inventor of the scale). Temperature on the Celsius scale (θ) is related to that on the thermodynamic scale (T) by

$$T=\theta+273.15 \qquad 5.1$$

The term degree Centigrade is no longer used, having been replaced by degree Celsius in 1948.

The measurement of temperature

We can seldom measure the temperature of an object directly. We must either add something to or take something from the object to measure its temperature. Usually we add a thermometer to the system under study and, after the new system has come into equilibrium, we measure the temperature of the thermometer. According to the zeroth law of thermodynamics, this will also be the temperature of the system. It will not, of course, be the temperature of the system before the thermometer was added.

A thermometer is simply an instrument in which the change in some physical property in response to temperature can be measured. This property may be the volume of a liquid or gas, the electrical resistance of a material or one of many other proprties. Each type of thermometer will have its own temperature scale and therefore must be calibrated against a standard. Normally, calibration consists of measurements at the defining or secondary fixed points of IPTS-68 together with a comparison against a standard platinum resistance thermometer, either directly, or more usually via an intermediate standard or standards (Barber, 1971; Collier, 1982). It is only practical to calibrate at, say, a maximum of 10 points, and so intermediate values are interpolated.

Errors of measurement

Because the measurement of temperature is indirect, there are two categories of error. First, there is the error in the measurement of the physical property. It may be a measurement of length, volume, resistance or voltage and will have some uncertainty depending on the method of measurement. The likely errors will differ for each type of thermometer, and will be considered when each type is discussed individually.

The second type of error depends on calibration. Each thermometer must be calibrated against a standard and this calibration will add its uncertainty to that of the standard. This source of error is often forgotten, but is frequently as large as the uncertainty of measurement of the physical property. We will use the term *accuracy* to denote how close our measurements are to the actual value of the temperature on the IPTS, and *precision* to indicate how closely our measurements made at one time agree with one another quite independently of any systematic error involved. *Stability* denotes the change in accuracy with time.

It is important to remember that temperature is defined only for a body in thermal equilibrium. Thus all the parts of the thermometer must be at the same temperature. This is rarely true in practice, except perhaps for total immersion thermometers. The errors arising from heat flow down the stem of a liquid-in-glass thermometer or along the connecting wires of a thermocouple or resistance thermometer are hard to estimate because they depend on many factors. Nevertheless, they can be considerable and should always be minimised. Partial immersion thermometers, however, are calibrated under standard conditions which include heat flow down the stem. Ideally, they should be used in the same conditions as those prevailing during the calibration.

The final problem is to identify which temperature we are measuring. In fact, we always measure the temperature of the thermometer. If it is in good contact and therefore in thermal equilibrium with the system under study, then this is also the temperature of the system. The system often includes more than we intend, when, for instance, radiation contributes significantly or there are air currents near a poorly attached thermometer. What we cannot measure, however, is the temperature of the system before we included the thermometer. Adding an ordinary laboratory thermometer to 100 cm^3 of water can easily change the temperature of the water by 0.1 K. Liquids of lesser heat capacity (such as organic solvents) will show a greater change, as will gases. If possible, it is preferable to install the thermometer as a permanent part of the system.

Techniques and instruments

Because any property that changes with temperature can be used as the basis of a thermometer, it is not surprising that many techniques are used for this purpose. Some techniques are better than others for particular jobs but there is a wide choice available for many measurements. We will restrict our discussion to those techniques likely to be of interest to biologists, that is, to the temperature range –20°C to 100°C (Table 5.1).

Table 5.1 The principal types of thermometer, their accuracy and features. Typical figures are quoted, but where appropriate the best performance achievable in a biological laboratory is given below the typical performance figures.

Type	Precision /K	Accuracy /K	Response time /s	Calibration interval	Advantages	Disadvantages
Liquid-in-glass	1	1	30 in water	5 years	–low cost –ease of use	–fragility –no electrical output
	0.01	0.02	1500 in water	frequent		
Platinum resistance	0.1	0.2	1 in water 10 in air	1 year	–stability –robustness –ease of use	–cost –size, but smaller units becoming available
	0.001	0.006		frequent		
Thermistor	0.1	0.3	5 in air	1 year	–low cost –ease of use –small size	–narrow temperature range of linear response
Band-gap device	0.2	1.0	15 in air	not known	–low cost –ease of use –robustness	–poor accuracy –self-heating
Thermocouple	0.1	0.5	1 in air	1 month	–low cost –ease of construction –small size	–many possible sources of error –need for reference temperature
Infra-red thermometer	0.1	1.5	5	5 years	–non-contact measurement	–cost –poor accuracy
LCD thermometer	3	1	1	not applicable	–low cost –surface measurement	–low precision –no electrical output

Mechanical thermometers

Almost all materials expand as they get warmer. In a perfect gas at constant pressure, the volume change is proportional to thermodynamic temperature. Gas thermometers using helium at low pressure come close to this ideal, with uncertainties of only a few mK for the best instruments (Guildner, 1982). However, gas thermometers are complex instruments suitable only for the standards laboratory.

The differential volume change of a liquid contained in a solid has provided the most common instrument for temperature measurement for over three centuries. The liquid is usually mercury, but sometimes, for temperatures below its freezing point, an alcohol may be used. It is contained in a glass bulb attached to a capillary tube. As the liquid expands or contracts relative to the volume of the bulb the length of the liquid column in the capillary lengthens or shortens.

Table 5.2 A selection of typical thermometers, their precision, accuracy and cost. The last is given as approximately the logarithm to the base 10 of UK price in £ in 1982.

Type	Precision /K	Accuracy /K	Manufacturer or supplier	Cost
Mercury-in-glass				
1. GP105c/0.5/100 (BS 1704)	0.5	0.6	Brannan	0.7
2. A40c/100 (BS 593)	0.1	0.4	Brannan	1
3. SR6/34c (BS 1900)	0.05	0.1	Zeal	2
Bimetallic				
4. THT-650-030A	1	1	Gallenkamp	1
Quartz				
5. 2804A	0.0001	0.04	Hewlett-Packard	3
Platinum resistance				
6. 80 ETI-2	0.1	0.2	Rosemount	2.3
7. 2189A	0.01	0.1	Fluke	3
8. 8520A/PRT	0.001	0.01	Fluke	4
Thermistor				
9. 4706	0.1	0.25	Digitron	2
Thermocouple				
10. 1624	0.1	0.5	Comark	2.5
11. 2190A	0.1	0.3	Fluke	3
Infra-red				
12. KT 14	0.25	1.5	Heimann	3
Fluoroptic				
13. 1000A	0.1	0.5	Luxtron	3.5
LCD contact				
14. 770-2	5	1	Telatemp	0
Thermal paint				
15. RP40-RP170	10	2	Thermographics	1.5 ($0.25\ dm^3$)

For laboratory use, this type of thermometer has no equal. It is cheap, relatively robust, and accurate if carefully used and periodically recalibrated (e.g. 1,2, and 3 in Table 5.2). Precision varies from 1 K for general purpose thermometers to better than 0.02

K for secondary reference thermometers. Accuracies are similar to this – many thermometers have an accuracy about equal to the precision of measurement. Many mercury-in-glass thermometers are built to a British standard (either BS 1704, BS 593 or BS 1900: British Standards Institution, 1951, 1974, 1976) and these should be referred to for full details of accuracy and precision.

Mercury-in-glass thermometers have several disadvantages. They are fragile and their principle of operation does not allow any simple method of electrical read-out for input to a data-logger. Their large mass is also a problem and means that they are used mainly for measuring the temperatures of liquids. The response time in air can be several minutes unless the air is forced over the thermometer bulb at speed. Thermal contact with solids is difficult, because these thermometers are usually cylindrical.

Another mechanical thermometer in common use is the bimetallic strip, in which the differential expansion of two strips of different metals produces a bending moment in a bar. This makes a particularly robust instrument but with a response time of minutes, poor accuracy, typically ±0.5 K (4), and no electrical read-out.

A third type of mechanical thermometer is of interest. Quartz crystals have resonant frequencies which are stable and repeatable functions of temperature. Accurate measurement of frequency is relatively simple electronically and the digital output is suitable for further processing by microcomputer. Each crystal must be individually calibrated, but this can be done under computer control and the calibration curve stored digitally. Accuracy is very good, ±40 mK from –5°C to 150°C, and precision approaching 0.1 mK allows very small temperature differences to be detected. This is an instrument of high quality, and carries a price to match (5).

Electrical thermometry

Several electrical properies of materials change with temperature. Many laboratory and field measurements are now made automatically by data-logger and electrical thermometers can be interfaced directly into these. This reduces both the cost and the inaccuracy of the measurement and is a major reason for the increasing popularity of electrical methods **(see Table 5.3)**.

Resistance thermometers. The electrical resistance of many materials varies with temperature but only two main classes of material are used in practical thermometers.

Certain metals have resistances which are very nearly linear functions of temperature. Platinum is the most commonly used and *platinum resistance thermometers* are used widely for the accurate measurement of temperature. They are made in many configurations but each contains a sensing element of high-purity, strain-free platinum wire. There may be two, three or four connections to this element depending on the increasing accuracy required. The resistance of the element can be determined by passing a known, small current through the sensor and measuring the voltage across the resistance. More accurate instruments incorporate a Wheatstone bridge circuit to reduce the current and hence self-heating.

Almost all commercially available platinum resistance thermometers have a resistance R of 100 Ω at 0°C and conform to DIN 43 760 (see also British Standard 1904: British Standards Institution, 1964). This specifies interchangeability within 0.1 Ω at 0°C, corresponding to an accuracy of 0.3 K. Many manufacturers offer better accuracy and specify this as 1/2 DIN, 1/5 DIN etc. Although the temperature coefficient, $R^{-1}(dR/dT)$, is small (0.0036 K^{-1} between 10°C and 30°C) the accuracy with which resistance can be measured means that the overall accuracies range from ±0.3 K at 0°C to ±0.6 K at 100°C for standard devices (e.g. 6). These accuracies can be improved further by calibration, and by the use of a precision bridge to measure resistance (7,8). Best accuracies achieved in standards laboratories are about ±2 mK between 0°C and 100°C (Guildner, 1982) and about ±50 mK should be attainable by most laboratories with generally available but expensive equipment (8).

Response times of platinum resistance thermometers vary considerably depending on size and construction but can be as little as 4 s in moving air for the latest type of evaporated thin-film elements. Stability is excellent with many manufacturers quoting ±0.05 K per annum, or better.

The *thermistor* is also a resistance thermometer but the material is a semi-conducting metal oxide mixture with a highly non-linear temperature response. Temperature coefficients are large, as much as 0.05 K^{-1}, making measurements of the change in resistance very easy. Thermistors are made in many forms with perhaps the most useful being glass or epoxy-coated beads ranging from 0.2 to 3 mm diameter. Small diameter beads have response times in still air of as little as 1 s. More robust types are also available, some encapsulated in stainless steel for particularly harsh environments.

The non-linear temperature response of thermistors means that the conversion from resistance to temperature is more difficult than with platinum resistance thermometers. Some manufactures provide equations relating resistance to temperature and a microcomputer can easily be programmed to make the conversion. Alternatively, simple resistor networks can be used to linearise the change in resistance over a restricted range of temperature (circuits can be found in many manufacturers' leaflets e.g. those of Fenwal Electronics (1974) and Thermometrics (1981)). This is adequate over a range of ±20 K about a selected temperature and some manufacturers supply fully encapsulated thermistor/resistor networks. By incorporating two thermistors in such a network the output can be made linear over a wider range of temperature (up to ±50 K) although at the expense of size and response time.

The interchangeability of thermistors has been a major problem. It has been difficult to manufacture thermistors of the same resistance although temperature coefficients have been closely matched. Some manufacturers now claim to have solved this problem and offer thermistors with an interchangeability of 0.2 K or better. Accuracy and stability are good, being largely dependent on the circuitry used for the measurement of the resistance of the thermistor. Precision thermistors are now available which change from their calibration by as little as 10 mK in 200 d if kept below 100°C (La Mers *et al.*, 1982). Cheap laboratory instruments offer overall accuracies

Table 5.3 Temperature sensors suitable for use in an automatic data-logging system.

Type	Precision	Accuracy	Response time	Temperature coefficient	Manufacturer or supplier
Platinum resistance thermometer	0.01 Ω	0.1 Ω	3 s	$4\,10^{-3}\,K^{-1}$	Digitron, 1754/2
Thermistor	5 mK	0.2 K	10 s	$4\,10^{-2}\,K^{-1}$	Fenwal, UUB31J1
Band-gap sensor	0.05 K	1.0 K	12 s	$1.0\,\mu A\,K^{-1}$	Analog Devices, AD590M
Thermocouple (type T)	0.1 K	0.4 K	1 s	$4\,10^{-5}\,V\,K^{-1}$	

Table 5.4 Temperature references for use with thermocouples or for calibrating thermometers. Cost as for Table 5.2.

Type	Stability /K	Accuracy /K	Manufacturer	Cost
Ice point reference	0.01	0.1	Mectron	2.5
Ice point reference	0.002	0.05	Delristor	3.0
Ceramic phase-change reference	0.05	?	Galai (Polarisers Technical Products)	2.5
Electronic ice point for thermocouples	1.0	1.0	Ancom	1.5

of ±0.3 K or better (e.g. 9). This will not be much improved by calibration because the inaccuracy is due largely to inadequacies in the circuitry for measuring resistance. Most manufacturers who want to improve significantly the accuracy of their resistance measurements turn to platinum resistance thermometers and thereby gain a wider measuring range as well as increased accuracy.

Semiconductor sensors. The voltage at which a semiconducting diode starts to conduct (called the band-gap of the junction) is an exponential function of temperature. This effect can be used by measuring the voltage across a forward-biased diode at constant current. The diode is then acting as a thermistor and will probably have a similar accuracy and stability when individually calibrated. Self-heating may be a problem because few small diodes have good thermal contact between the semiconductor die and the case.

The change in band-gap with temperature has also been used in an integrated circuit to produce an output which is a linear function of temperature. The devices can be made interchangeable by laser trimming the appropriate resistors during manufacture. At present, accuracy is rather low, ±1.7 K from –55°C to 150°C, but a single point calibration can improve this to better than ±1 K. In air, the response time is about 10 s but self-heating can add up to 0.7 K to the estimated temperature. These devices are therefore only suitable for measuring the temperature of liquids or solids, but the attraction of a device producing 1 $\mu A\ K^{-1}$ independent of supply voltage fluctuations is considerable.

Thermocouples. Until recently the thermocouple was the universal temperature sensor in the biological laboratory in applications for which the mercury-in-glass thermometer was unsuitable. This device works on the principle that, if two wires of different metals are joined at each end and the two junctions are at different temperatures, there will be a current in the wires depending on the temperature difference. If one of the wires is cut, the potential difference across the cut ends when kept at the same temperature will be proportional to the temperature difference – the Seebeck effect. This potential difference is small, usually less than 1 mV for a temperature difference less than 20 K.

Much has been written about the correct use of thermocouples – see, for example, Perrier (1971), James (1980), Reed (1982), British Standard 1041 Part 4 (British Standards Institution, 1966) and the many leaflets and handbooks published by thermocouple manufacturers. When correctly used, the thermocouple is highly accurate, capable of measuring temperature differences of less than 1 mK. However, this accuracy is only attained by the most stringent control of temperature in every part of the measurement circuit, the complete exclusion of interference and the use of sophisticated instruments to measure the minute potential difference. The normal laboratory thermocouple is capable of nothing near this performance. Errors arise from electrolytic effects at the junction, from spurious thermal potentials caused by thermal gradients in the wires and in the microvoltmeter, from electrical interference and from mechanical stress in the wires and switches.

Thermocouple accuracy is difficult to specify. Standard thermocouples are specified to ±0.8 K in British Standard 4937 (British Standards Institution, 1973,1974), but thermocouples with half this uncertainty are also available commercially. Calibration can improve this figure but stability then becomes the main problem. High precision thermocouple thermometers (e.g. 11) have specified accuracies of about ±0.2 K at calibration, widening to ±0.3 K after one year (type T thermocouple). Normal laboratory instruments are usually specified to have an accuracy of about ±0.5 K after calibration (e.g. 10).

Thermocouples have a number of advantages over other electrical thermometers. The measuring junction can be very small and hence of low thermal capacity. Response times of less than 1 s in still air are easily attained. They are cheap, although accurate measurements require a good, and hence expensive, microvoltmeter.

Thermocouples do not measure absolute temperature but temperature difference between the two junctions. If absolute temperatures are required then one junction must be kept at a known, fixed temperature (commonly the ice point) and this adds to the uncertainty in the measurement. If, however, temperature differences are required (as in psychrometers and heat flow meters) then thermocouples are the obvious choice. Electronic references are now commonly incorporated into thermocouple thermometers, but have poor accuracy (e.g. ±1 K) although they are very convenient in portable instruments.

Optical thermometry

Radiation can carry information about the temperature of an object. Measurements of temperature can then be made without contact, reducing disturbance to the system and the problem of energy transfer by conduction and convection associated with contact devices.

Infra-red thermometers. Every object radiates energy depending on its temperature. If the object is a black-body (that is, it has a perfectly emitting surface with an emissivity ϵ of unity), then the spectrum of that radiant energy is given by Planck's radiation equation

$$L_\lambda = 2c^2h/\lambda^5(\exp(hc/\lambda kT)\text{-}1) \qquad 5.2$$

in which L_λ is the spectral radiance at wavelength λ, h is Planck's constant, c is the velocity of light, k is Boltzmann's constant and T is the thermodynamic temperature. Near T equal to 300 K, this distribution has a broad peak at about 10 μm in the infra-red part of the spectrum (Fig. 5.1). At the same wavelength, objects at different temperatures will have different spectral radiances, given by Planck's equation multiplied by the emissivities of their surfaces, i.e. by ϵL_λ.

Many surfaces have emissivities which peak between 0.9 and 1 at about 8-10 μm wavelength, so that radiance in this range provides a useful measure of temperature. For example, white paper, white lacquer and polished glass have emissivities of 0.93, 0.95, and 0.94 respectively. Clean or polished surfaces have much lower emissivities, as low as 0.02 for gold (values from Oriel Infra-red Products Catalogue and Weast, 1972).

Fortuitously there is an atmospheric window from 8-12 μm where the transmissivity of radiation in air is high. Absorption bands of water and other constituents of the air strongly absorb radiation at wavelengths either side of this window (Fig. 5.1). Thus, to estimate temperatures from 0°C to 500°C, many infra-red thermometers measure radiance in the waveband 8-12 μm minimising errors due to the detection of atmospheric radiation. At higher temperatures, other wavelengths, such as the 3-4 μm window, are used.

Various sensors detect infra-red radiation. Thermal detectors respond to changes in temperature caused by the absorption of radiation. Thermistor bolometers are basically sensitive thermistors and thermopiles are stacks of thermocouples in series. Pyroelectric crystals, such as lithium tantalate, respond to infra-red radiation by producing a voltage proportional to the change in temperature. Thermal detectors are rather slow because of their large thermal mass. Most respond in ms although pyroelectric crystals can operate at frequencies up to 2 kHz.

Faster and more sensitive infra-red photovoltaic and photoconductive semiconductor detectors are available but few will detect at wavelengths longer than 8 μm. The compound semiconductors PbSnTe and HgCdTe are most sensitive near 10 μm but

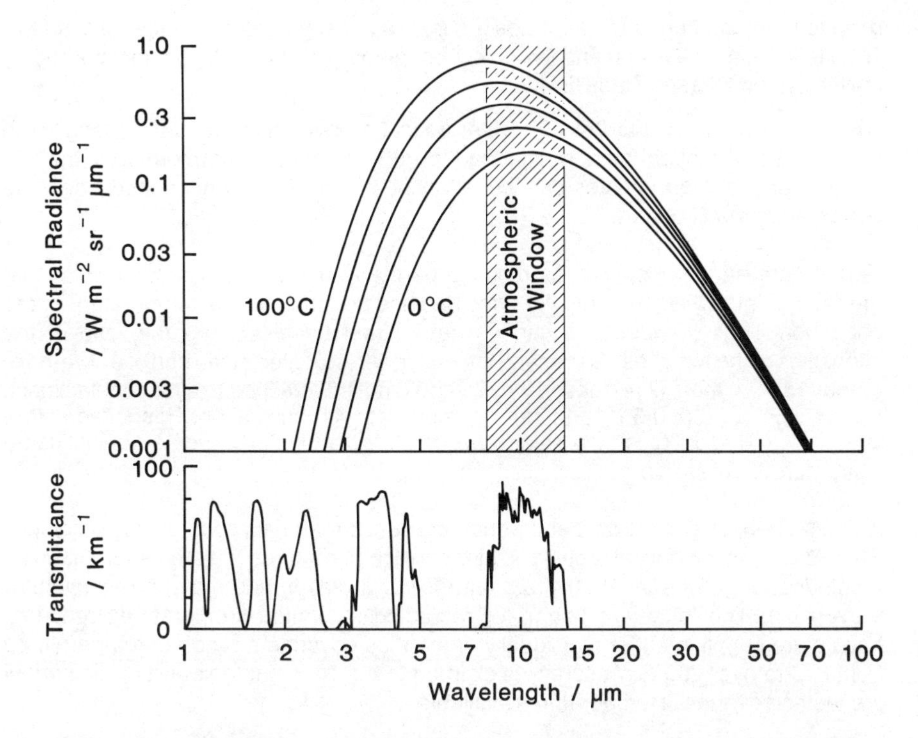

Figure 5.1 Black-body spectral radiation curves for temperatures of 0°C, 25°C, 50°C, 75°C and 100°C showing the atmospheric window used by many infra-red thermometers. The lower graph is the percentage transmission of infra-red radiation over 1 km in the wavelength range 1-15 µm.

must be operated at liquid nitrogen temperature. It is unlikely that they will be used in portable infra-red thermometers in the near future.

Thermal detectors have low sensitivity (of the order 10 V W^{-1}) and detect only changes from ambient temperature. In precision instruments they are therefore used with a rotating shutter and alternately view the object whose temperature is being measured and the shutter or radiation reflected by the shutter. A phase-locked amplifier amplifies the minute a.c. signal representing the temperature difference between the object and the reference. The temperature of the body is computed from the difference signal and the temperature of the reference as measured by a thermistor or resistance thermometer within the case of the infra-red thermometer.

Such infra-red thermometers are sophisticated pieces of electro-optical engineering. Cheaper instruments are also available which measure the d.c. signal from a bolometer without the use of a reference. These must be adjusted frequently by reference to a black-body source at known temperature, sometimes built into the instrument itself. Such instruments usually have an uncertainty of several kelvins, but even the best infra-red thermometers are limited by the detection technique and, despite many refinements in the design, their accuracy is no better than ±1 K (e.g. 12). Response time is fast and the object under investigation may be undisturbed by the measurement (see Table 5.4).

Other optical thermometers. Other optical properties of materials change with temperature. Although these do not lead to true non-contact thermometers, they do mean that the thermometer can be attached to the body under study and the temperature read from a distance.

Certain rare-earth phosphors fluoresce at particular wavelengths in such a way that the ratio of intensities in particular spectral lines depends on the temperature of the phosphor. Thus, temperature can be determined by measuring the ratio of the intensities of the two lines. At present, the phosphor is located at the end of a fibre-optic probe (13), but there is no reason why it should not be attached to a moveable object. Accuracy is good (±0.5 K) and can be improved with calibration. Response time is fast (1 s) and stability is also said to be good. Unfortunately, such a sophisticated instrument is expensive.

A much cheaper type of optical thermometer uses materials such as liquid crystals which change their optical properties at precise temperatures. This is usually seen as a colour change of a label (14), a paint (15) or a crayon and may be reversible or irreversible. The latter are useful for recording maximum temperatures reached. Labels are supplied either individually or in strips covering a range of temperatures from 0°C to over 250°C. Accuracy is claimed to be ±1 K but their utility is limited by the restricted range of temperatures available.

Calibration

It is a basic principle of measurement that what an instrument measures should be traceable either to a fundamental physical property or to another instrument of greater accuracy than the first. In thermometry, general laboratory and field thermometers are calibrated against a secondary reference mercury-in-glass thermometer and this in turn is calibrated against a platinum resistance thermometer which has been calibrated at the fixed points of IPTS-68. Barber (1971) discusses in detail the calibration of thermometers and the errors involved.

Calibration is normally carried out by immersing both thermometers in a well-stirred fluid bath held at constant temperature. If the thermal capacity of the bath is large the inevitable cycling of the bath temperature as the thermostat switches on and off is unimportant. However, this ensures that it will take time to change the bath temperature so that calibration over a range of temperatures will be slow. A regression of the response of the test thermometer (y) against temperature measured on the standard thermometer (x) then gives the calibration relation.

Two types of error are estimated from the calibration. Firstly, the random error of measurement, which is the precision of the test thermometer, is estimated from the regression as the root mean square of the unaccountable variance. Secondly, the accuracy is derived from the limits to the systematic uncertainty of the regression. In regression analysis, these are usually given as the standard errors in the prediction of y given x, or as confidence limits of the regression line. However, we are interested in the inverse problem, that of estimating *x* (the temperature as measured by the standard thermometer) given *y* (the reading of the test thermometer).

If the deviations are small and the slope of the line is nearly unity, as they should be for a calibration, then the standard error of prediction of *x* on *y* is approximately equal to that of *y* on *x* within the range of both variates (see Sprent, 1969, pp 97-99). Thus the usual formula for the standard error of prediction can be used to estimate the uncertainty of the calibration of the test thermometer. This will be largest at the extremes of the range of calibration and least near the centre. Ideally, it will be small everywhere in the range of interest compared with the random error of measurement, and if not, more measurements should be taken until it is. The estimated errors rise rapidly at the extremes of the range of calibration, so that thermometers should be calibrated over a wider range than that in which they will be used.

Various organisations including the British Standards Institution and, for more accurate thermometers, the National Physical Laboratory, calibrate thermometers. The British Standards Institution recommends calibration at 5-yearly intervals for liquid-in-glass thermometers, with more frequent checks at a fixed point on the scale.

Many types of thermometer have an accurately reproducible temperature response but vary in absolute calibration. They need frequent (say 3-monthly) calibration at one temperature to maintain accuracy. This is usually done at the ice point (0°C at normal atmospheric pressure), the normal boiling point of water (100°C) or the triple point of water (0.01°C). The first is the most easily attained because no control of pressure is needed, the temperature change with pressure being so small, -0.07 K MPa^{-1}, that it can be neglected.

Most types of thermometer can be calibrated in a fluid bath but infra-red thermometers need a source of radiation at known temperature. This will be a black-body source, best provided by a cavity radiator – a block of metal at uniform temperature containing a cavity with an exit hole just large enough to admit the optical head of the infra-red thermometer (Barber, 1971). The inner surface of the cavity should be diffusely reflecting with an emissivity exceeding 0.7 for the radiation in the cavity to approach closely that of a black-body at the temperature of the metal.

Black-body radiators are supplied by several manufacturers and the temperature of the metal is usually measured with a platinum resistance thermometer or an accurate mercury-in-glass thermometer. One manufacturer offers a novel device where the cavity temperature is controlled by a phase transition in a ceramic. Stability of ±0.05 K after only 5 minutes operation is claimed, although the absolute accuracy is not quoted.

Some manufacturers also offer calibrators for resistance thermometers and thermocouples. These are precision resistance or voltage standards, sometimes marked in units of temperature via the appropriate conversions. They are used to calibrate the resistance and voltage measuring circuits associated with resistance thermometers and thermocouples and cannot be used instead of a calibration of the thermometer itself.

Case studies

1 Leaf temperature

Temperature is an important variable in most studies on leaves yet it is particularly difficult to measure. Most leaves are planar, hence the thermometer must be pressed against one surface, but both surfaces are actively gaining or losing heat by radiation, convection and transpiration. Hence the thermometer must be small, in practice comparable with or smaller than the thickness of the boundary layer of air next to the leaf. It must not significantly affect this boundary layer or convection and transpiration will be disrupted, and it must not cover a significant area of leaf or radiation will be changed also.

Only two temperature sensors approach these requirements. Miniature bead thermistors and thermocouples are small enough (<0.5 mm) to be used pressed against the underside of the leaf. However, the large flux of heat mostly absorbed at the top surface of the leaf and lost at the lower implies a temperature difference across the boundary layer. The leads of the temperature sensor will tend to 'short-circuit' this and cause errors in temperature measurement. The sensor itself will distort the air flow and hence change the heat flow near the sensor. The large flux of heat through the leaf also means it is not in equilibrium, and the top surface of the leaf may be significantly hotter than the lower surface if the irradiance is high. Any temperature measurement will be of the lower (partly shaded) surface of the leaf and this will be less than the temperature averaged across the leaf. All the above errors can be significant and make the estimation of leaf temperature very uncertain. Claims of accuracy of better than ± 1 K should be treated with caution.

An infra-red thermometer is a better instrument for measuring leaf temperature if the source of radiation illuminating the leaves has little or no infra-red component. If the radiation source radiates infra-red radiation in the waveband 8-12 μm, much of this radiation will be reflected off the leaf and some will be transmitted through the leaf. This reflected and transmitted infra-red radiation will seriously interfere with the measurement.

2 Microclimate measurements

Many animal and plant physiologists wish to monitor the environment of their organisms, and air temperature is an important variable in the environment. Furthermore, micrometeorologists can derive the fluxes moving from crop to atmosphere or *vice versa* from measurements at various heights in the crop.

Almost any type of accurate thermometer can be used but must be adequately ventilated. In the field, natural wind velocities are often high enough to do this but forced ventilation will be necessary in confined spaces or sheltered environments close to the ground. The thermometer must also be shielded from radiation without restricting air flow – see case study 3.

For micrometeorological applications, temperature gradients are important. Biscoe *et al.* (1978) used a ventilated thermistor above the crop and a differential thermocouple array within the crop to measure the temperature gradient. Although the thermocouple junctions were not shielded from radiation, the authors claim that the corrections for radiation and convective heat exchange are a second order correction in the temperature difference over a short height interval in the crop.

Fast response and high precision are needed for eddy correlation measurements, for which Verma *et al.* (1979) used microbead thermistors and fine-wire (25 μm diameter) chromel-constantan thermocouples. They obtained good correlations between temperature spectra from the two sensors, but above 30 °C the thermistors registered average temperatures about 0.2-0.4 K higher than did the thermocouples, because of errors in the calibration procedure. More recently, B.J. Legg has used a resistance thermometer 0.5 mm long and 0.6 μm diameter to measure temperature fluctuations of less than 0.01 K at frequencies up to 3 kHz.

3 Air temperature – the effect of exposure

We have discussed the errors in temperature measurement that can be ascribed to the choice and calibration of thermometers. However, in the field, the errors associated with thermometer exposure may be an order of magnitude greater than the calibration errors. Fritschen and Gay (1979) discuss the errors which arise from ventilation, conduction along the leads, radiation and time-lag. The last two are usually important only in air, because of the poor coupling between a thermometer and the atmosphere compared to that in water, the volumetric specific heat of water being over 3000 times that of dry air at STP. We are indebted to Dr J.A. Clark of the University of Nottingham School of Agriculture for his help with this case study.

The radiation error of a thermometer, $\triangle T$, the difference between the temperature of the thermometer and the temperature of the air, is given by

$$\triangle T = R/h \qquad 5.3$$

where R W m^{-2} is the average heat gain from net radiation over the thermometer surface and h W m^{-2} K^{-1} is the coefficient of convective heat transfer between the thermometer and its surroundings. Over the range of wind speed relevant to micrometeorology, it can be shown (Monteith, 1973, pp. 101, 102 and 224) that

$$h \simeq 4(u/d)^{0.5} \qquad 5.4$$

for a cylindrical thermometer of diameter d m in a wind speed of u m s^{-1}. The following examples illustrate the possible sizes of radiation errors in measurements of air temperature and their changes with ventilation, thermometer size and radiation, estimated from these two equations.

Ventilation. Assume $R = 100$ W m^{-2} and $d = 5$ mm

u/m s^{-1}	$\triangle T$/K
0.2	4.0
0.5	2.5
1.0	1.8
2.0	1.3
5.0	0.8

Size. Assume $R = 100\ \mathrm{W\ m^{-2}}$ and $u = 1\ \mathrm{m\ s^{-1}}$

d/mm	ΔT/K
5.0	1.8
1.0	0.8
0.5	0.6
0.1	0.3
0.01	0.1

Radiation. Assume $d = 5$ mm and $u = 1\ \mathrm{m\ s^{-1}}$

R/W m^{-2}	ΔT/K
300	5.3
100	1.8
30	0.5
10	0.2

These calculations show that radiation errors can be substantial even at the ventilation rates usually recommended for thermometers ($u > 3\ \mathrm{m\ s^{-1}}$) and for thermometers that are physically small. In strong sunshine, radiation errors can be decreased to levels comparable to those incurred in calibration only by minimising radiation exchange by using reflective shields and reflective thermometer surfaces. Long (1957) used a hemicylinder of copper gauze painted matt white to shield his nickel resistance thermometers and found that this was adequate in sunshine as long as the wind speed exceeded 1 m s^{-1}. Radiation errors may also be significant on still, clear nights, and the principle of least work suggests that they warrant as much attention as calibration errors.

4 Animal core temperatures

Animal core temperatures are often measured by inserting a thermometer into a convenient external opening in the animal. For long-term monitoring of free-ranging animals this is impossible and some form of telemetry must be used. Temperature transducer/transmitters can be surgically implanted or fed to the animal, but must be small and hermetically sealed. Thermistors are probably the most useful sensor, as their high sensitivity reduces the complexity and hence size of the circuitry to encode temperature into the transmitted output. Boland and Bell (1980) used bead thermistors in an epoxy-coated package force-fed to young crocodiles to monitor their core temperatures. The thermistor resistance determined the period of the pulses produced by the transmitter which could be monitored at distances up to 30 m from the animal.

5 Soil temperatures

To measure soil temperature, thermometers and their electrical connections must be strengthened and protected to avoid damage from soil movement, abrasion and corrosion from the soil solution.

Many climatological studies use mercury-in-glass thermometers. To 200 mm below the surface, bent-stem thermometers are permanently installed so that the strengthened glass bulb and stem below ground are vertical and the graduated scale is horizontal at the soil surface. Those which sense at or deeper than 300 mm hang in air by a chain from the cap of a vertical steel tube, with the bulb at the specified depth. The thermometer is fitted into a stout outer glass sheath with the bulb embedded in paraffin wax to ensure correct readings when hauled to the surface to be read.

For his micrometeorological studies, Long (1957) potted nickel resistance thermometers in resin and encased them and the leads in pliable polyethylene tubing. This protected them when they were installed in horizontal holes bored into the exposed face of a pit dug into the soil and subsequently back filled.

6 Temperature integration

Some ecological and phenological studies require integrated temperature. This can be realised in several ways.

The response of a mercury-in-glass thermometer is damped by surrounding its bulb with a mass of water in a glass container. Its temperature then fluctuates with a reduced amplitude and lags in time compared with air temperature fluctuations, but approaches the mean air temperature after a sufficiently long period. This is the basis of the temperature integrating bottle of Winspear and Morris (1959), used to estimate average overnight temperature in glasshouses.

The rate at which a chemical reaction proceeds depends on temperature, and this can be used to estimate mean temperature. The rate of hydrolysis of sucrose has been used extensively in ecology (Jones, 1972) for this purpose, particularly in central Europe.

Signals from electrical thermometers can be averaged electronically to measure integrated temperature. Buckley (1979) has described circuitry to integrate temperature above certain thresholds to monitor the progress to maturity of a crop in terms of degree-days. Green *et al.* (1983) describe a semiconductor band-gap temperature sensor for temperature integration in the field.

Summary

It is easy to be uncritical about the measurement of temperature. The human hand can detect temperature differences of less than 1 K but few general thermometers have an accuracy as good as this. Biochemical processes with a Q_{10} of 2 change their rate by 7.2% for every 1 K change in temperature yet the thermostats on the water baths in which we perform our reactions are rarely accurate to better than 1 K. Simple animals can detect temperature differences of less than 1 K and the distribution of plants may be determined by differences in climate of less than this.

Temperature can be measured accurately only by careful attention to the measurement process. The thermometer (of whatever type) must have been calibrated accurately, it must be used in accordance with the manufacturer's instructions and all possible sources of error must be considered and where possible minimised. Readings should be repeated (replicated) and the possibility of systematic error recognised. Many experimenters will carefully randomise their experimental treatments but never suspect that their thermometer may show a systematic drift in performance during the day as it is used and as ambient temperature changes. This should be allowed for in experimental design.

No measurement is complete without a statement of error. This is particularly important for temperature measurement, as even expensive thermometers are accurate to only 1 part in $3\ 10^3$ (0.1 K in 300 K), yet many other laboratory instruments can measure far more accurately (1 part in $2\ 10^5$ for a voltmeter of comparable cost). Even the fixed points on the temperature scale are specified to only 1 part in $3\ 10^4$. Despite this, many temperature measurements are reported without error limits. Many thermometers are accurate only to about ± 1 K, so that quoting temperatures to tenths of a kelvin is often misleading. Many false claims of accuracy appear, encouraged by widely optimistic claims by some manufacturers. This situation should improve as the market matures and some of the less technically sophisticated suppliers disappear.

Bibliography

There are many texts both in physics and biology which describe the effects of temperature but there is little published about how to measure it. The classic work is the multi-volume "Temperature: its measurement and control in science and industry" which has been published every decade since 1941, and most recently in 1982 (Schooley, 1982). Basic principles of temperature measurement are covered in many elementary physics texts such as Halliday and Resnick (1978) but these have few practical details. However the recent book by Nicholas and White (1982) should fill this gap.

Much useful information can be gleaned from the various leaflets and handbooks published by thermometer manufacturers. The following list is not exhaustive but each item contains useful information about thermometry and may usually be obtained free of charge from the company shown.

Calex Instrumentation. Temperature Handbook. Leighton Buzzard, Bedfordshire, England. (This is an extract from the Omega Temperature Measurement Handbook mentioned below).

Fenwal Electronics (1974). Thermistor Manual. Framingham, Massachusetts, U.S.A. (distributed by Electroautom Ltd., Maidstone, Kent, England).

Labfacility Ltd. (1982). Temperature Sensing with Thermocouples and Resistance Thermometers. Hampton, Middlesex, England. (Price £1.50).

Linseis GMBH (1979). Handbook of Temperature Measurements. Selb, West Germany (U.K. office in Grays, Essex, England).

NANMAC Corporation (1979). Temperature Handbook. Framingham, Massachusetts, U.S.A.

Omega Engineering Inc. (1981). Temperature Measurement Handbook. Stamford, Connecticut, U.S.A.

Oriel Corporation. Infra-red Products. Stamford, Connecticut, U.S.A. (U.K. office in Kingston-upon-Thames, Surrey, England).

CGS/Thermodynamics (1974). The Thermocouple Handbook. Southampton, Pennsylvania, U.S.A. (distributed by International Instruments Ltd., Hounslow, Middlesex, England).

Thermometrics Inc. (1981). Thermistors. Cat. No. 181-A. Edison, New Jersey, U.S.A.

Articles comparing various thermometers also appear occasionally in the trade press in such magazines as Electronics Industry and Laboratory Equipment Digest (for example that by Dance, 1979).

References

Barber, C.R. (1971). The Calibration of Thermometers. London: HMSO.

Biscoe, P.V., Clark, J.A., Gregson, K., McGowan, M., Monteith, J.L. & Scott, R.K. (1975). Barley and its environment I. Theory and practice. J. appl. Ecol., **12**, 227-257.

Boland, J.E. & Bell, C.J. (1980). A radiotelemetric study of heating and cooling rates in unrestrained, captive *Crocodylus porosus.* Physiol. Zool., **53**, 270-283.

British Standards Institution (1951). British Standard 1704: General purpose thermometers.

British Standards Institution (1964). British Standard 1904: Industrial platinum resistance thermometer elements.

British Standards Institution (1966). British Standard 1041: Part 4: Thermocouples.

British Standards Institution (1973, 1974). British Standard 4937: Parts 1 to 7: International thermocouple reference tables.

British Standards Institution (1974). British Standard 593: Laboratory thermometers.

British Standards Institution (1976). British Standard 1900: Specification for secondary reference thermometers.

Buckley, D.J. (1979). A precision, multi-threshold, multi-readout temperature integrator. Agric. Meteorol., **20**, 1-6.

Collier, R.D. (1982). Calibration with confidence: the assurance of temperature accuracy. *In* Temperature: Its Measurement and Control in Science and Industry. Volume 5, ed. J.F. Schooley, pp. 1311-1315. New York: American Institute of Physics.

Dance, M. (1979). MPUs allow high precision thermometry. Electron. Ind., **5**(5), 12-17.

Fritschen, L.J. & Gay, L.W. (1979). Environmental Instrumentation. New York: Springer-Verlag.

Green, C.F., Schaare, P.N. & Bates, C.N. (1983). A temperature sensor for temperature integration in the field. J. exp. Bot., **34**, 226-229.

Guildner, L.A. (1982). The measurement of thermodynamic temperature. Phys. Today, **35**(12), 24-31.

Halliday, D. & Resnick, R. (1978). Physics. Third edition. New York: Wiley.

James, R.W. (1980). Thermocouple developments: pitfalls with precision. Lab. Equip. Dig., **18**(3), 95-97.

Jones, R.J.A. (1972). The measurement of mean temperatures by the sucrose inversion method. A review. Soils Fertil., **35**, 615-619.

La Mers, T.H., Zurbuchen, J.M. & Trolander, H. (1982). Enhanced stability in precision interchangeable thermistors. *In* Temperature: Its Measurement and Control in Science and Industry. Volume 5, ed. J.F. Schooley, pp. 865-873. New York: American Institute of Physics.

Long, I.F. (1957). Instruments for micro-meteorology. Q. Jl. R. met. Soc., **83**, 202-214.

Monteith, J.L. (1973). Principles of Environmental Physics. London: Arnold.

National Physical Laboratory (1976). The International Practical Temperature Scale of 1968: amended edition of 1975. London: HMSO.

Nicholas, J.V. & White, D.R. (1982). Traceable Temperatures. Wellington, N.Z.: DSIR.

Perrier, A. (1971). Leaf temperature measurement. *In* Plant Photosynthetic Production: Manual of Methods, eds. Z. Sestak, J. Catsky & P.G. Jarvis, pp. 632-671. The Hague: Dr W. Junk.

Reed, R.P. (1982). Thermoelectric thermometry: a functional model. *In* Temperature: Its Measurement and Control in Science and Industry. Volume 5, ed. J.F. Schooley, pp. 915-922. New York: American Institute of Physics.

Schooley, J.F. (1982). Temperature: Its Measurement and Control in Science and Industry. Volume 5. New York: American Institute of Physics.

Sprent, P. (1969). Models in Regression and Related Topics. London: Methuen.

Verma, S.B., Motha, R.P. & Rosenberg, N.J. (1979). A comparison of temperature fluctuations measured by a microbead thermistor and a fine wire thermocouple over a crop surface. Agric. Meteorol., **20**, 281-289.

Weast, R.C. (1972). Handbook of Chemistry and Physics. 53rd edition. Cleveland, Ohio: Chemical Rubber Co.

Winspear, K.W. & Morris, L.G. (1959). Temperature integrating bottles for measuring mean air temperature. J. agric. Engng. Res., **4**, 214-221.

Chapter 6: The measurement of wind speed

J. Grace
Department of Forestry and Natural Resources,
University of Edinburgh, King's Buildings,
Mayfield Road, Edinburgh, EH9 3JU, Scotland.

Introduction

Developments in aeronautics, engineering and meteorology provide a wide range of sensors from which to choose. It should be borne in mind that a practical measurement of wind speed is usually the sum of several interactions between the air flow and the response of the sensor. Consequently different types of sensor measure rather different attributes of the wind.

The purpose of this chapter is to discuss the general principles and designs of anemometers, bringing up-to-date the reviews of Caborn (1968), Mazarella (1972), Grace (1977) and Fritschen and Gay (1979). It is not possible in a short space to discuss turbulence: suitable background reading is Bradshaw (1971) and Raupach and Thom (1981). I have not attempted to discuss windvanes, feeling unable to improve on Fritschen and Gay (1979).

In choosing an anemometer, the following criteria are among the most important to consider:

1) The range of wind speed over which the sensor operates. Many mechanical anemometers have a starting threshold wind speed, below which they do not operate.

2) The *linearity of response* is crucial if integration over a period of time is required. Some sensors are inherently non-linear, but can be used with a lineariser circuit to enable integration. Otherwise non-linear responses must be integrated using a microcomputer in which the calibration curve is stored as part of a program.

3) The speed of response of many anemometers is dependent on wind speed. Hence the *time constant*, normally defined as the time taken for a sensor, exposed to a step-wise change in conditions, to reach 63% of its complete response, is not very useful. The more useful, and analogous *distance constant* is the distance of air that must pass the sensor to achieve 63% of the complete response.

4) The size of the instrument is often crucial when working in the environment of vegetation, so is the mechanical endurance.

5) Wind speed is a vector quantity, with attributes of magnitude and direction. The wind vector U may be resolved on x, y and z Cartesian co-ordinates, resulting in u, v and w components of wind speed (Fig. 6.1). If the aim is to measure unambiguously any one of these, then the sensor must have a cosine response and also indicate the sign.

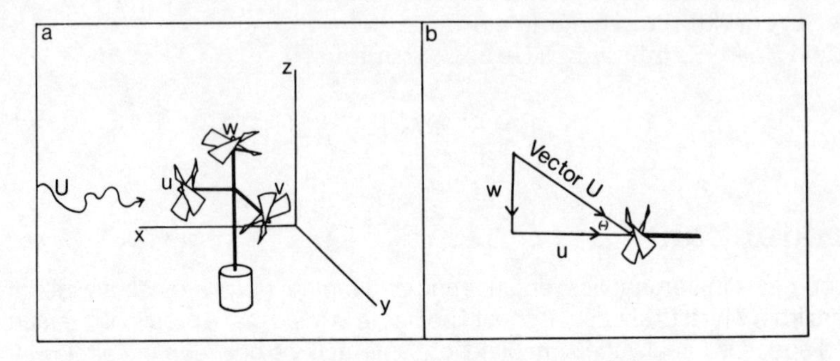

Figure 6.1 a) Resolution of the wind vector into u, v and w components by a hypothetical set of sensors pointing into the wind, across the wind, and across and vertical to the wind. b) The cosine response required to measure the components shown in a: when the wind vector U makes an angle θ with the main axis of the sensor, the signal from the sensor should be $U\cos\theta$.

Pressure probes

Pitot-tube

When a tube connected to one arm of a manometer is pointed into the wind the manometer reads a pressure which can be related to wind speed. A simple open tube to measure the dynamic pressure is not sufficient since the reading on the manometer would have to be corrected for the prevailing static pressure which would be sensed by a measuring device at rest relative to the fluid. Thus the manometer is used in a differential mode to sense the difference between the static pressure P_s and the dynamic pressure P_d:

$$\triangle P = P_d - P_s = 0.5\rho u^2 \qquad 6.1$$

An early arrangement employed two tubes. The first one pointed into the wind, sensing P_d. The second terminated in a streamlined plug and had holes in its walls to sense P_s. The modern instrument combines both tubes in a concentric arrangement, the Pitot-static tube (Fig. 6.2).

The advantages of the Pitot-static tube as a standard are that its characteristics have been exhaustively determined (Ower and Pankhurst, 1966; Bryer and Pankhurst, 1971), and that it is very easy to use at least at high wind speeds [1]. It does not require especially careful orientation into the wind: 12° errors in alignment produce virtually no change in pressure sensed.

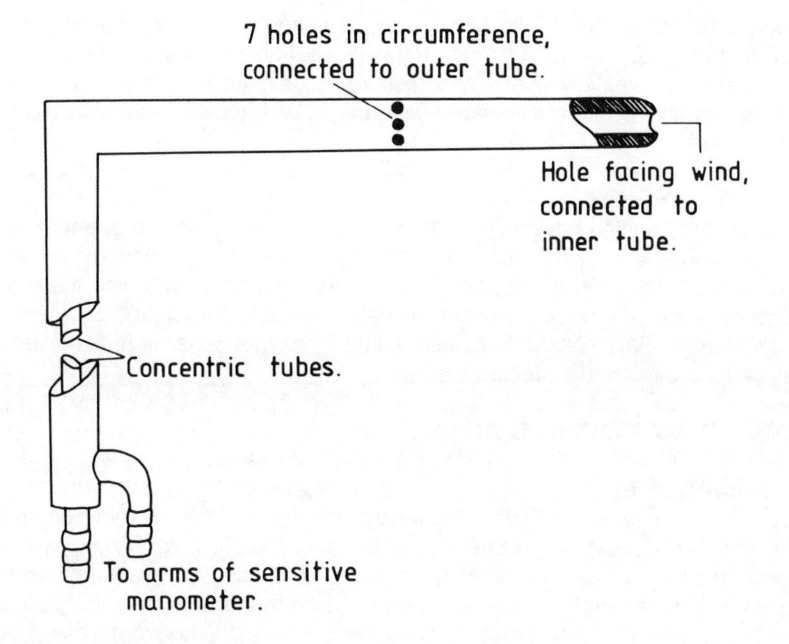

Figure 6.2 Pitot-static tube

The range of wind speeds of interest in environmental physiology is however very low in relation to developments in the field of aeronautics. Ordinary manometers in conjunction with Pitot tubes are not useful when working below about 5 m s^{-1}. Highly sensitive manometers in which the liquid meniscus is sighted through a microscope, and in which the manometer tube is set on an adjustable, gentle incline instead of being held vertical, increase the usefulness of the Pitot-static tube greatly. The National Physical Laboratory (NPL) design, the Combist micromanometer [4], will detect 0.02 Pa, equivalent to a wind speed of 0.2 m s^{-1}. More sensitive still are electronic manometers, which incorporate various types of pressure transducer (Bryer and Pankhurst, 1971). Some of these are specifically designed for use with a Pitot tube, and are available with a "square root extractor" to give a signal proportional to wind speed [7]. Such systems do of course need calibrating against liquid manometers, but as they give a continuous electrical signal they are very useful [17].

Great care is required when using Pitot-static tubes below 1 m s^{-1}, however, as the pressure response to velocity is influenced by poorly understood viscous effects: Bryer and Pankhurst (1971) show how an appropriate correction can be made, based on the internal diameter of the Pitot tube in question. It is probably wise to employ alternative calibration methods below 1 m s^{-1}.

Pressure tube anemograph

The pressure tube anemograph is based on the same principle as that of the Pitot tube, but designed for installation 10 m above the ground at major meteorological stations

(Meteorological Office 1956). It is mounted on a suitable wind vane so that it always points into the wind. The standard instrument [8] will not measure wind below 2 m s^{-1}, being limited by a relatively insensitive pressure sensing and recording device. Similar instruments, with more sensitive recording equipment, have sometimes been used in shelter research (Jensen, 1954).

Pressure – sphere

Other pressure sensing devices have been designed for use in micrometeorological work. Thurtell *et al.* (1970) describe a spherical sensor with several ports. The pressure difference between various sets of holes can be used to find the u, v and w components of the wind. A rather similar instrument, used in conjunction with a rapid response temperature sensor, has been used to measure sensible heat flux by eddy correlation (Tanner and Thurtell, 1970).

Sensors activated by drag

Cup anemometer

The cup anemometer is widely used to measure the mean of the horizontal component of wind speed (Fig. 6.3). It consists of a set of cups, usually 3, connected by arms to a vertical spindle. Its main advantage is that the rotation rate is a linear function of wind speed over a wide range, and so the mean can be found by simply integrating the signal over a period of time. Being truly omnidirectional in the horizontal plane it does not have to be mounted on a windvane. It must however be mounted with the spindle vertical: a 10° misalignment can cause a 6% error in the measurement of the horizontal component of wind speed.

Figure 6.3 Cup anemometer set, showing 3-cup and 6-cup rotors (Vector Instruments, Rhyll).

The original form of the instrument (Patterson, 1926) overestimated a fluctuating wind speed, but subsequent modifications to the shape of the cup almost eliminated this problem (Sheppard, 1940; Scrase and Sheppard, 1944; Deacon, 1951; Rider, 1960 and Fritschen, 1967). Most recent papers on errors in using cup anemometers are concerned with the influence of masts and the errors caused by imperfect cosine response in the vertical plane (Moses and Daubek, 1961; MacCready, 1966; Gill *et al.* 1967; Bernstein, 1967; Izumi and Barad, 1970 and Kaganov and Yaglom, 1976).

One of the main limitations of cup anemometers is that they do not begin to rotate until a threshold windspeed is attained. This speed varies with the design and age of the instrument but is usually in the range 0.1 – 0.4 m s^{-1}. Thus, substantial errors are likely to occur in the integration of wind speed over any time interval during which such low wind speeds occur. Another limitation is evident when the instrument is used in a highly-turbulent flow, such as behind a windbreak. The anemometer sometimes rotates in reverse and measures nothing that can be related to the processes of heat and mass transfer. For this reason, the use of cup anemometers inside vegetation is unwise, even though some designs are small enough for this purpose (Bradley, 1969).

For field surveys over long periods, robust versions with mechanical counters are useful [8]. The more sensitive types provide an electrical output, either as an analogue voltage or as a train of pulses [3,12,15]. Very low inertia units with cups of expanded polystyrene in sets of 6 or 3 are available [12]. They have a lower threshold starting speed and a distance constant of only 1 m instead of 5 m in the standard cups (Jones, 1965).

Despite the difficulties just outlined, determination of wind profiles over vegetation is generally accomplished with a set of six cup anemometers. Calibration must be very carefully done as small errors in the mean wind speed can cause surprisingly large errors in the derived quantities, especially in the roughness length z_o (Tanner, 1963). The temperature gradient must be measured at the same time, so that a stability correction can be incorporated where necessary (Paulson, 1970; Dyer, 1974; Thom, 1975; Businger, 1975; Yaglom, 1977).

Propeller or vane anemometer

The original vane anemometer was developed in England a hundred years ago and has been sold by one company since then [18]. It is essentially a windmill, made of flat metallic vanes, with some means of measuring the rotation (Ower and Pankhurst, 1966). Several commercial versions are available, designed mainly for measuring unidirectional flow in ducts or pipes (Fig 6.4a). This type of instrument is particularly useful in testing growth cabinets, fume cupboards and the flow through ventilation fans. It responds in a practically linear manner to wind speed, and in some versions has a very low starting threshold (0.15 m s^{-1}).

The propeller anemometer most frequently used in micrometeorological work is Gill's design (Gill, 1975). This has a 4-blade helicoid-shaped propeller, made of expanded polystyrene, which has low inertia and better cosine response than the flat-bladed metal version. The original model was mounted on a lightweight bi-vane which pointed the propeller into the wind, aligning to both azimuth and elevation angle. The

Figure 6.4 Ultra-sensitive vane anemometer (Lowne Instruments, London). On the left is shown the Leda 1000 sensing head with average windspeed display. On the right, a triad of sensors being used to measure *u*, *v* and *w* components of the wind. (Photographs by Lowne Instruments and Dr. G. Bent).

calibration coefficient is determined by the pitch of the propeller (which is fixed) and the sensitivity of the tachometer-generator. The development of this single-propeller model, leading to a 3 propeller model capable of measuring *u*, *v* and *w* components of wind speed [13] simultaneously is described by Gill (1975).

The Gill anemometer, because of its low inertia, has often been used for eddy correlation measurements (Dyer *et al.*, 1967; Hicks *et al.*, 1975; Thompson, 1979; Milne, 1979). It does not respond to very high frequency fluctuations and thus underestimates the heat or mass transfer (McBean, 1972; Hicks, 1972; Garratt, 1975; Dyer *et al.*, 1982). Its distance constant, at the most favourable angles, is 1 m, but at oblique angles becomes much greater (Garratt, 1975). It has been much criticised on account of its poor cosine response, although correction for this is to some extent possible (Horst, 1973). The calibration becomes non-linear below 1 m s^{-1}, due to friction in the bearings, a point of considerable concern when investigating the *w* component which is often less than this.

Very recently an 8-bladed version of the flat vane anemometer, with digital electronic output, has been used in micrometeorological work by the Ecological Physics Research Group at Cranfield Institute in Bedfordshire (Figs. 6.4a and 6.4b). It has several advantages over the Gill anemometer: a practically perfect cosine response, a 0.25 m distance constant, a threshold of 0.11 m s^{-1}, excellent linearity and small size (Schaefer and Bent, pers. comm.). This anemometer is likely to gain wider acceptance in the future.

Pressure plate and thrust anemometer

The earliest known anemometer was a swinging plate, free to be deflected in the wind. Versions used in the present century are described by Rees (1927), Sherlock and Stout (1931) and Jensen (1954). There is some recent work towards the development of a plate anemometer in which the plate is rigidly held and small elastic distortions are sensed with strain gauges (Krause and Fralick, 1980). Similar equipment has been used in micrometeorological work to measure the *w* component of wind (Redford, Verma and Rosenberg, 1981).

The thrust anemometer described by Smith (1980) consists of a sphere on a vertical stalk. The stalk pivots on a fulcrum half-way along its length. Wind forces on the sphere deflect the rod from the vertical, and its motion is detected by displacement transducers housed at the base of the rod. Direction, as well as speed is obtained.

Probes depending on convective heat loss

Hot-wire anemometer

The hot-wire anemometer depends on heat transfer from a small wire element to the surrounding fluid. It is a well-established sensor used widely in engineering studies (Comte-Bellot, 1976; Freymuth, 1978; Perry, 1982).

In its simplest form current is passed through a very fine wire to bring it to a temperature considerably above the ambient. When air flows across this sensor, the latter is cooled, and the consequent change in electrical resistance can be detected as a change in the voltage across the wire. A calibration curve relating wind speed and voltage can be established. The sensor is incorporated into a Wheatstone bridge to permit measurement of small changes in resistance (Fig. 6.5a). This arrangement is referred to as a constant current anemometer, although it does not run at constant current but constant voltage.

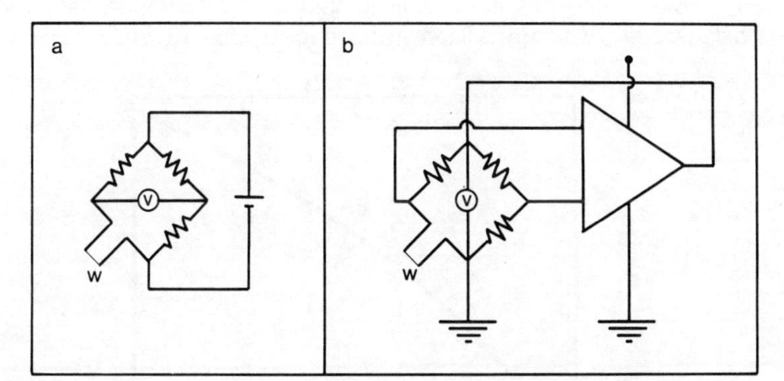

Figure 6.5 Hot wire anemometer circuits. a) Wheatstone bridge, with wire *w* as a constant voltage anemometer b) constant temperature mode: when the bridge is unbalanced the servo amplifier increases the bridge top voltage, thus maintaining the wire at a constant resistance and temperature.

However, such a system suffers from the disadvantage that the probe-to-air temperature difference declines with increasing air speed and, indeed the response is so non-linear that it becomes worthless at wind speeds in excess of a few metres per second. The useful range of the sensor is increased when it is used in the constant temperature mode (Fig. 6.5b). When air flows across the sensor the change in electrical resistance unbalances the bridge and the error voltage activates the servo amplifier. This applies a voltage at the top of the bridge, heating the sensor until its resistance and therefore its temperature is restored to the original value and the bridge is balanced. An additional advantage of the constant temperature mode is its more rapid response time, extending to several kHz, essential in some types of study where high frequency turbulence is of interest.

The relationship between wind speed and heat loss from an infinite cylinder was studied by King (1914 a and b) and can be derived from standard relationships given in text books of heat transfer (Kreith, 1965). One form of King's Law, relevant to the constant temperature anemometer, relates the voltage output V to the wind speed u as follows

$$V^2 = a + bu^n \tag{6.2}$$

where a and b are constants for a given wire, and n is typically 0.5. This markedly non-linear relationship between V and u is especially unsatisfactory when attempting to describe the properties of a rapidly fluctuating wind, as an above-average reading of wind will be recorded as a much smaller change in V than a below-average reading. Consequently, research anemometers incorporate a linearising circuit, designed to produce a voltage proportional to velocity [5,9,10].

There is, however, a further source of non-linearity. At very low wind speeds, King's Law (or any other equation based on forced convection relationships) does not apply because natural convection replaces forced convection as the dominant mode of heat transfer (Fig. 6.6). Even this can be accommodated by commercially available systems, at least one of which [5] is linear in the range 0.025 – 10 m s^{-1}.

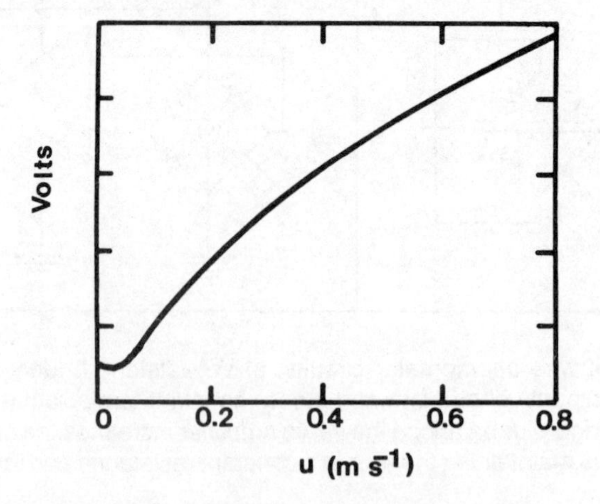

Figure 6.6 Typical calibration of a hot wire anemometer.

A typical hot wire probe is made of 5-25 μm diameter platinum, and is 1-3 mm long (Fig. 6.7). The small size ensures small thermal mass and hence a rapid response. The wire is mounted between fine prongs, tapering to a needle point to minimise heat loss by conduction. Perry (1982) describes the manufacture of such probes from Wollaston wire [12], though many designs are commercially available. The most complex probes utilize the directional sensitivity of the wire, in a 3-wire device for analysis of *u*, *v* and *w* components of turbulence. Although hot-wire anemometers have occasionally been used in eddy correlation (Bottemanne, 1979) they are not ideal for this, being fundamentally unable to distinguish an updraught from a downdraught.

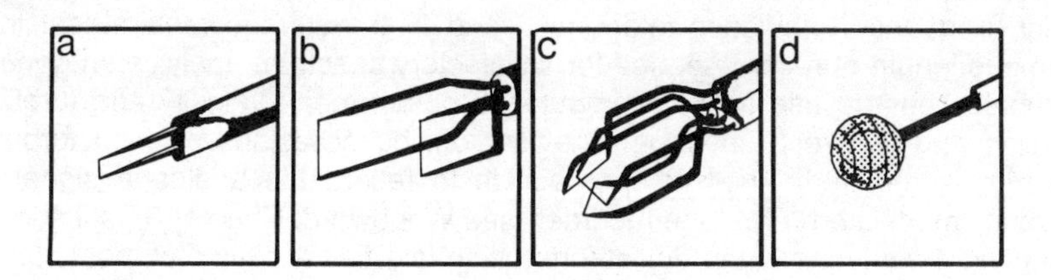

Figure 6.7 Configuration of hot wire anemometer probes: a) single wire b) two-wire probe in which second wire provides temperature compensation c) triple-wire sensor for resolution of directional components d) omnidirectional film-type sensor.

Although sometimes used in the field (Larsen and Busch, 1976), all these probes are fragile and best used indoors in a clean environment. Typical applications are Grace and Wilson (1976) and Raupach *et al.* (1980). In field conditions, some protection is necessary (Finnigan, 1979).

More rugged probes, with similar characteristics to hot-wires, have been marketed recently. The film probes are made by depositing a thin layer of metal on a quartz substrate and covering it with a protective layer of ceramic. A potentially very useful design is the spherical probe which responds nearly equally to wind from all directions (Lang and Leuning, 1981).

Probes like these but rugged enough for industrial use are also commercially available.

In the constant temperature anemometer, the response time of the instrument is not limited by the thermal inertia of the wire and hence the output is practically a measure of the instantaneous velocity. Thus, the passage of an individual eddy over the wire results in a rapid fluctuation in the output voltage. Rapid fluctuations may be explored in a qualitative way by simply displaying the voltage on an oscilloscope screen, and runs of several seconds can be recorded using a polaroid camera. The statistical properties of these fluctuations which collectively make up what is generally called turbulence can only be investigated with more elaborate equipment.

The equipment generally required for the quantitative analysis of such signals includes an integrating digital voltmeter, a true mean square voltmeter and a spectrum analyser. The first two of these are relatively cheap and enable evaluation of only the most basic summary statistics, the mean velocity and turbulence intensity. A spectrum analyser is an instrument which takes a stream of data over a time period, and then, either mathematically or by electronic filtering, displays the amplitude of the signal as a function of frequency. Spectrum analysers are extremely expensive to buy [6], although are readily available for hire over periods of a few weeks [14]. To some extent, these standard items of equipment can be replaced by a microcomputer equipped with a fast analogue to digital converter. The main limitations of the microcomputer approach are 1) the speed of data aquisition required for the capture of high frequency turbulence information and 2) the storage requirement for the enormous length of record needed for satisfactory spectrum analysis. In general, it is probably best to use a microcomputer to obtain mean velocity and turbulence intensity, and to hire a spectrum analyser on the occasions when a turbulence spectrum is required. Another approach is to record the analogue signal on a frequency modulated (FM) tape recorder (see Woodward, Chapter 8), subsequently translating the information to digital form using facilities available at most computer centres.

Related techniques

There are several related sensors which depend, like the hot-wire anemometer, on the cooling of hot objects, but differ in being relatively massive and so unable to detect rapid fluctuations.

The earliest of these, the Kata-thermometer, was used to assess the ventilation of rooms (Hill *et al.*, 1916). It consisted of a large bulbed thermometer which was heated in a water bath, dried with a cloth, and allowed to cool. Its modern equivalent is the omnidirectional probe referred to above (Ower and Pankhurst, 1966).

Numerous anemometers have been constructed in which the heated element is a thermistor. As with hot-wire anemometers, the simplest circuits are not to be recommended (Bergen, 1971; Unwin, 1980), as in these the principle of the heated element is not fully exploited since the probe is not run as a constant temperature device. Moreover, complicated corrections are required to allow for ambient temperature. In other instruments the sensor is a thermocouple wrapped around a heater (Fritschen and Shaw, 1961).

Commercially available instruments of this general type often have their sensors protected in a tubular sheath, and are mainly useful for checking air flow in growth cabinets and buildings. Usually, such systems produce a non-linear output. They are usually designed with temperature compensation circuitry and constructed ruggedly with the industrial market in mind [1,9,10,15].

Naphthalene – the basis for a mass transfer probe?

Naphthalene sublimation from a sphere or a plate proceeds at a rate which is a known function of windspeed and temperature (Neal, 1975). One disadvantage of this approach is that the function is not linear and there is thus no easy way to convert an observed weight loss to a mean windspeed. On the other hand, neither is heat loss from organisms a linear function of windspeed, and in studies where heat loss is the prime concern, it is arguably more valid to study mass transfer through the boundary layer, as an analogue for heat, than it is to measure windspeed *per se*. Moreover, there are problems in micrometeorology at the microscale which cannot be approached with any conventional sensor. The air flow between fur and feathers, or between cereal grains in a granary, are examples of this. Naphthalene can be deployed as very small granules or as a film spread over surfaces, and as such is likely to be a very sensitive detector of air movement. As a means of measuring the boundary layer resistance, it has the advantage over water in that its change of state is accompanied by only a small absorption of heat, and so the surface temperature remains practically constant and uniform in any experiment. Like water though, its vapour pressure is a strong function of temperature, and hence the temperature of the subliming surface must be known in order to evaluate the boundary layer resistance.

Anemometers measuring the flight time of a tracer

Pulsed wire anemometer

Although utilizing a heated wire, this anemometer works on an entirely different principle. Pulses of current passed along a central wire cause the release of heat pulses. These act as tracers and the elapsed time is measured between their release and detection by heat sensors on either side of the central wire. The sensor appears to be especially valuable in the analysis of highly turbulent flows behind bluff bodies (Bradbury and Castro, 1971; Tombach, 1973). Unlike a hot-wire anemometer it senses the direction of flow as well as the velocity. Its main limitation is that its calibration is not linear, and that its lower sensing range is limited by the tendency for natural convection to occur below about 0.25 m s^{-1}.

Vortex shedding anemometer

A smooth cylinder of diameter d, exposed to a wind speed u, sheds vortices at a frequency related to the Strouhal number S (Goldstein, 1957)

$$S = fd/u \qquad 6.3$$

where f is the frequency and S has a value of about 0.20. Thus, any fast sensor such as a hot wire, placed a suitable distance downwind from the cylinder can be used to measure the frequency of vortex-shedding, and hence enable calculation of the wind speed from the equation above. The equation holds over a limited range of Reynolds numbers, but by employing interchangable cylinders of varying d, a wide range of wind speeds can be measured. This principle is the basis for a type of industrial flowmeter [19] but has rarely been applied in anemometry.

Sonic anemometer

This instrument works on the principle that the speed of sound is increased if the air is moving in the same direction as the sound. Practical instruments measure the phase shift in sound waves travelling between two points (Fig. 6.8). Versions of the sonic anemometer have been described by Mitsuta (1966), Campbell and Unsworth (1979) and Shuttleworth *et al.* (1982). It has been developed commercially for eddy correlation, as it outperforms the Gill anemometer as far as linearity, sensitivity and frequency response are concerned. It can in addition, measure temperature. On the other hand, the cosine response has not been widely investigated, and in the design of Campbell and Unsworth (1979), was shown to be rather poor.

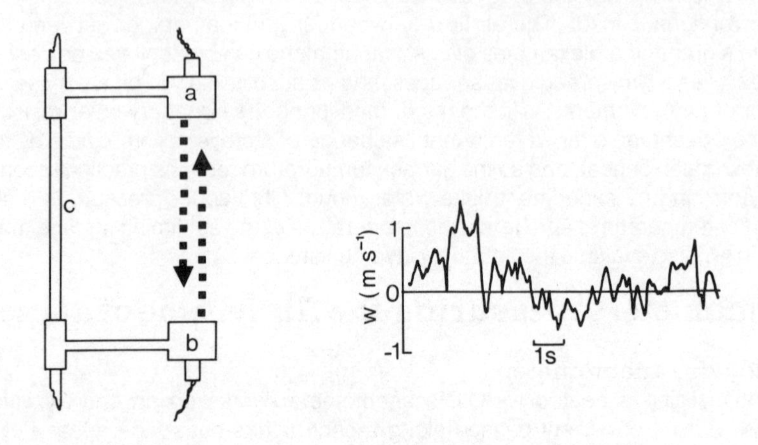

Figure 6.8 Sonic anemometer (Campbell and Unsworth, 1979). Ultrasonic transducers *a* and *b*, mounted on a mast *c* are used as both transmitters and receivers. The resulting record of the *w* component of the wind is shown on the right.

Laser-Doppler anemometer

The laser-Doppler anemometer is now widely used to measure flow in many industrial research applications where, to some extent, it has replaced the hot wire anemometer. A comprehensive bibliography of the subject was prepared by Durst and Zaré (1974). Descriptions of the principles of this measuring system are provided by Abbiss *et al.* (1974) and Durst *et al.* (1975).

The laser-Doppler anemometer depends on the detection of a Doppler shift in the frequency of radiation scattered by small particles which happen to be present in, or can be added to, the flow (Fig. 6.9). For velocities occuring near surfaces the Doppler shift is very small, but can readily be detected with appropriate optical systems and detecting units. There are several advantages over the hot wire anemometer. The laser and its associated equipment are bulky but operate at a fair distance from the flow being investigated and so have the advantage of not disturbing the flow at all. This is true of no other system. The relationship between output and velocity is linear and the anemometer does not require calibration. The spatial resolution is very fine, as the optics are focussed to 'look' at a tiny spot. It has distinct advantages over the hot wire anemometer when operating within a millimeter of a surface: the hot-wire overestimates windspeed in these circumstances, for reasons which are not properly understood (Hebber, 1980).

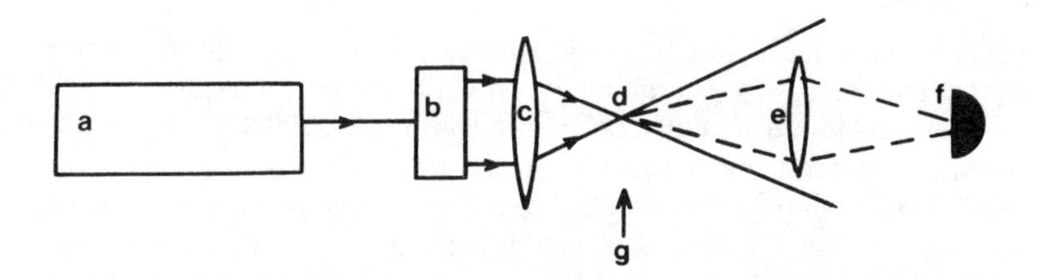

Figure 6.9 Laser-Doppler anemometer, used in the dual beam mode. A beam from the laser, *a*, is split into two equal beams and brought to a focus by the lens, *c*. At *d*, the two beams interact to make interference fringes. Particles in the flow, *g*, produce pulses of scattered light as they traverse these fringes. Scattered light is gathered by the lens, *e*, and the pulses are detected at *f*.

So far, the high cost of the system seems to be one of the main factors precluding its use, for laser anemometer systems are widely available in the U.K. [5,9,10].

Semi-quantitative methods

Flags

The use of flags to obtain a general impression of wind-speed over a large area was pioneered by the Forestry Commission (Lines and Howell, 1963). The rate of loss of area from standard flags, made of cotton, was supposed to integrate the wind speed over periods of weeks. The main limitation of this technique is that whereas the loss of flag area is a linear function of wind speed for dry flags, it is markedly non-linear when the flags are wet (Rutter, 1965).

Visualization

Movement of air can be traced using bubbles, preferably filled with a helium-air mixture resulting in an overall bubble density nearly equal to that of surrounding air so that the bubble neither rises nor falls but moves with the air (Moen, 1974). Commercially-available equipment [2] is designed to produce uniform, small bubbles a few mm in diameter. However, the smallest bubbles have significant inertia and so do not follow the smallest eddies. When suitably illuminated, the bubbles can be photographed and if the shutter speed is known the wind-speed can be estimated. Smoke generating equipment is also commercially available, and useful in wind tunnels. Among plants, smoke frequently reveals features about the flow which would otherwise go unnoticed (Oliver, 1973; Arkin and Perrier, 1974; Bergen, 1975).

In relation to individual leaves, Schlieren photography is useful in detecting the development of a thermal boundary layer especially in natural convection (Barnes and Bellinger, 1945).

Calibration

It is usual to calibrate anemometers in laminar flow in a wind-tunnel. For cup and propeller anemometers, fairly large wind tunnels are required (see Grace, 1977). Wind tunnels themselves are normally calibrated with a pitot-static tube connected to a sensitive manometer. In most wind tunnels the relationship between wind speed and fan speed is linear, and so after an initial calibration, it is subsequently necessary only to select specified fan speeds. Most wind tunnels do not run well below 1 m s^{-1} and the motor may stall at very low speed. Moreover, Pitot-tube calibration is not adequate in this range. Consequently, alternative calibration procedures are required.

The simplest way to calibrate anemometers below 1 m s^{-1} is to tow them on a trolley in a corridor, using a motor with gears to obtain a range of velocities. Calibration of hot wire probes to their lower detection limit poses a problem as most ordinary wind tunnels cannot run at such low speeds. Various miniature calibration rigs have been suggested. The simplest of these is made by connecting a glass pipe to the top of a large vessel of water. The water is allowed to drain from a hole near the bottom of the vessel, causing air to be drawn through the pipe at a flow rate proportional to the head of water. Other methods are discussed by Ower and Pankhurst (1966), Pratt and Bowsher (1978), Leutheusser and Aydin (1980) and Perry (1982).

Dynamic calibration, designed to test anemometer response to a fluctuating wind, is much more difficult. Perry (1982) provides a discussion of this, in relation to hot-wires.

Calibration factors may change over a prolonged period, especially when instruments with moving parts are kept in the field. Calibrations should be checked before and after use.

Influence of masts

The supporting structure to which the anemometer is attached inevitably disturbs the flow (Moses and Doubek, 1961; Gill *et al.*, 1967; Cermak and Horn, 1968). In micrometeorological work the sensors are lightweight and the mast can be quite slender (Wucknitz, 1977). It must however, be rigid, as any vibration will cause errors, especially in the high frequency range of turbulence measurements. Dyer (1981) presents visualization tests, which give some idea of the extent of flow distortion caused by struts.

Acknowledgements

I would like to thank those people who gave up their time to discuss parts of the original manuscript, especially Simon Allen, Andrew Sandford and Ronald Milne.

Manufacturers

1. Airflow Developments Ltd., Lancaster Road, High Wycombe, HP1 3QP, Buckinghamshire, England.

2. Armfield Engineering Ltd., Bridge House, West Street, Ringwood, BH24 IDX, Hampshire, England.

3. C.F. Casella and Co. Ltd., Regent House, Britannia Walk, London, N1 7LU, England.

4. Combustion Instruments Ltd., The Causeway, Staines, Middlesex, TW18 3AP, England.

5. Disa, Techno House, Redcliffe Way, Bristol BS1 6NU, England.

6. Honeywell Test Instruments, Charles Square, Bracknell RG12 1EB, England.

7. Mercury Electronics, Pollock Castle Estate, Newton Mearns, Glasgow G77 6NU, England.

8. R.W. Munro, Cline Road, Bounds Green, London N11 2LY, England.

9. Prosser Scientific Instruments, Lady Lane Industrial Estate, Hadleigh, Ipswich IP7 6DQ, England.

10. T.S.I., c/o Bristol Industrial and Research Associates Ltd., P.O. Box 2, Portishead, Bristol BS20 9JB, England.

11. Texcel Ltd., 13 Cunningham Hill Road, St. Albans, Herts., AL1 5BX, England.

12. Vector Instruments, 113 Marsh Road, Rhyl, Clwyd LL18 2AB, England.

13. R.M. Young Company, 2801 Aeropark Drive, Traverse City, Michigan 49684, U.S.A.

14. Livingstone Hire Ltd., Shirley House, Camden Road, London NW1 9NR, England.

15. Didcot Instrument Company Ltd., Station Road, Abingdon, Oxon OX14 3LD, England.

16. Light Laboratories, 10 Princes Street, Brighton BN2 1RD, England.

17. Chell Instruments Ltd., Tudor House, Grammar School Road, North Walsham, Norfolk NR28 9JH, England.

18. Lowne Instruments Ltd., Boone Street, London SE13 5SA, England.

19. J.Tec Associates, Unit 5, Moniton Industrial Estate, Worthing Road, Basingstoke RG22 6NQ, England.

20. Technitron (England.), Doman Road, Yorktown Industrial Estate, Camberley, Surrey GU15 3DM, England.

21. Johnstone Matthey Metals, 100 High Street, Southgate, London N14 16ET, England.

References

Abbiss, J.B., Chubb, T.W. & Pike, E.R. (1974). Laser Doppler anemometry. Opt. & Laser Technol., 1974, 249-261.

Arkin, G.F. & Perrier, E.R. (1974). Vorticular air flow within an open row crop canopy. Agric. Meteorol., **13**, 359-374.

Barnes, N.F. & Bellinger, S.L. (1945). Schlieren and shadowgraph equipment for air flow analysis. J. opt. Soc. Am., **35**, 497-509.

Bergen, J.D. (1971). An inexpensive heated thermistor anemometer. Agric. Meteorol., **8**, 395-405.

Bergen, J.D. (1975). Air movement in a forest clearing as indicated by smoke drift. Agric. Metorol., **15**, 165-179.

Bernstein, A.B. (1967). A note on the use of cup anemometers in wind profile experiments. J. appl. Meteorol., **6**, 280-286.

Bottemanne, F.A. (1979). Eddy correlation measurements above a maize crop using a simple cruciform hot-wire anemometer. Agric. Meteorol., **20**, 397-410.

Bradbury, L.J.S. & Castro, I.P.C. (1971). A pulsed wire technique for velocity measurements in highly turbulent flows. J. Fluid Mech., **49**, 657-691.

Bradley, E.F. (1969). A small, sensitive anemometer system for agricultural meteorology. Agric. Meteorol., **6**, 185-193.

Bradshaw, P. (1971). An introduction to turbulence and its measurement. Oxford: Pergamon.

Bryer, D.W. & Pankhurst, R.C. (1971). Pressure-probe methods for determining wind speed and flow direction. National Physical Laboratory. London: HMSO.

Businger, J.A. (1975). Aerodynamics of vegetated surfaces. *In* Heat and Mass Transfer in the Biosphere, ed. D.A. de Vries and N.H. Afgan, pp. 139-165. New York: Wiley.

Caborn, J.M. (1968). The measurement of wind speed and direction in ecological studies. *In* The Measurement of Environmental Factors in Terrestrial Ecology, 8th Symposium of the British Ecological Society, pp. 69-81. Oxford: Blackwells.

Campbell, G.S. & Unsworth, M.H. (1979). An inexpensive sonic anemometer for eddy correlation. J. appl. Meteorol., **18**, 1027-1077.

Compte-Bellot, G. (1976). Hot-wire anemometry. Ann. Rev. Fluid Mech., **8**, 209-231.

Cormack, J.E. & Horn, J.D. (1968). Tower shadow effect. J. geophys. Res., **73**, 1869-1876.

Deacon, E.L. (1951). The over-estimation error of cup anemometers in fluctuating winds. J. scient. Instrum., **28**, 231-234.

Durst, F. & Zaré, M. (1974). Bibliography of Laser-Doppler-Anemometry Literature. Unpublished manuscript.

Durst, F., Zaré, M. & Wigley, G. (1975). Laser doppler anemometry and its application to flow investigations related to the environment of vegetation. Boundary-Layer Meteorol., **8**, 281-322.

Dyer, A.J. (1974). A review of flux-profile relationships. Boundary-Layer Meteorol., **7**, 363-372.

Dyer, A.J. (1981). Flow distortion by supporting structures. Boundary-Layer Meteorol., **20**, 243-251.

Dyer, A.J., Hicks, B.B. & King, K.K. (1967). The Fluxatron – a revised approach to the measurement of eddy fluxes in the lower atmosphere. J. appl. Meteorol., **6**, 408-413.

Dyer, A.J. & 21 others (1982). An international turbulence comparison experiment (ITCE 1976). Boundary-Layer Meteorol., **24**, 181-209.

Finnigan, J.J. (1979). Turbulence in waving wheat. I Mean statistics and Honomi. Boundary-Layer Meteorol., **16**, 181-211.

Freymuth, P. (1978). A bibliography of thermal anemometry. TSI Quarterly, **4**(4),3-26. St. Paul, MN 55164, USA: TSI Incorporated.

Fritschen, L.T. (1967). A sensitive cup-type anemometer. J. appl. Meteorol., **6**, 695-698.

Fritschen, L.J. & Gay, L.W. (1979). Environmental Instrumentation. New York, Heidelberg, Berlin: Springer-Verlag.

Fritschen, L.T. & Shaw, R.H. (1961). A thermocouple-type anemometer and its use. Bull. Am. met. Soc., **42**, 42-46.

Garratt, J.R. (1975). Limitations of the eddy-correlation techniques for the determination of turbulent fluxes near the surface. Boundary-Layer Meteorol., **8**, 255-259.

Gill, G.C. (1975). Development and use of the Gill UVW Anemometer. Boundary-Layer Meteorol., **8**, 475-495.

Gill, G.C., Olsson, L.E., Sela, J.I. & Suda , M. (1967). Accuracy of wind measurements on towers or stacks. Bull. Am. Met. Soc., **48**, 665-674.

Goldstein, S. (1957). Modern Developments in fluid dynamics. Oxford: Clarendon.

Grace, J. (1977). Plant Response to Wind. London: Academic Press.

Grace, J. (1983). Plant-Atmosphere Relationships. London: Chapman and Hall.

Grace, J. & Wilson, J. (1976). The boundary layer over a Populus leaf. J. exp. Bot., **27**, 231-241.

Hicks, B.B. (1972). Propeller anemometers as sensors of atmospheric turbulence. Boundary-Layer Meteorol., **3**, 214-228.

Hicks, B.B., Hyson, P. & Moore, C.J. (1975). A study of eddy fluxes over a forest. J. appl. Meteorol., **14**, 58-66.

Hill, L., Griffith, O.W. & Flack, F. (1916). The measurement of the rate of heat loss at body temperature by convection, radiation, and evaporation.

Horst, T.W. (1973). Corrections for response errors in a three-component propeller anemometer. J. appl. Meteorol., **12**, 716-725.

Izumi, Y. & Barad, M.L. (1970). Wind speeds as measured by cup and sonic anemometers and influenced by tower structure. J. appl. Meteorol., **9**, 851-856.

Jensen, J.M. (1954). Shelter effect. Copenhagen: Danish Technical Press.

Jones, J.I.P. (1965). A portable sensitive anemometer with proportional d.c. output and a matching velocity-component resolver. J. scient. Instrum., **42**, 414-417.

Kaganov, E.I. & Yaglom, A.M. (1976). Errors in wind-speed measurements by rotation anemometers. Boundary-Layer Meteorol., **10**, 15-34.

King, L.V. (1914a). On the convection of heat from small cylinders in a stream of fluid: determination of the convection constants of small platinum wires with application to hot-wire anemometry. Proc. R. Soc. Lond. Ser. B., **90**, 563-570.

King, L.V. (1914b). On the convection of heat from small cylinders in a stream of fluid. Phil. Trans. R. Soc. Lond. Ser. A., **214**, 273-432.

Krause, L.N. & Fralick, G.C. (1980). Miniature drag force anemometer. Rept. No. NASA-TM-81680 E-706, National Aeronautics and Space Administration, Cleveland, Ohio, USA.

Kreith, F. (1965). Principles of Heat Transfer. New York: International Textbook Company.

Lang, A.R.G. & Lenning, R. (1981). New omnidirectional anemometer with no moving parts. Boundary-Layer Meteorol., **20**, 445-457.

Larsen, S.E. & Busch, N.E. (1976). Hot-wire measurements in the atmosphere. Disa Inf. (Den.), **20**, 5-21.

Leutheusser, H.J. & Aydin, M. (1980). Very low velocity calibration and application of hot-wire probes. Disa Inf. (Den.), **25**, 17-18.

Lines, R. & Howell, R.S. (1963). The use of flags to estimate relative exposure of trial plantations. Forestry Commission, Forest Rec., Lond., **51**, 1-31.

McBean, G.A. (1972). Instrument requirements for eddy correlation measurements. J. appl. Meteorol., **11**, 1078-1084.

MacCready, P.B. (1966). Mean wind speed measurements in turbulence. J. appl. Meteorol., **5**, 219-225.

Mazzarella, D.A. (1972). An inventory of specifications for wind measuring instruments. Bull. Am. met. Soc., **53**, 860-871.

Milne, R. (1979). Water loss and canopy resistance of a young Sitka spruce plantation. Boundary-Layer Meteorol., **16**, 67-81.

Mitsuta, Y. (1966). Sonic anemometer-thermometer for general use. J. met. Soc. Japan, **44**, 12-24.

Moen, A.N. (1974). Turbulence and the visualization of wind flow. Ecology, **55**, 1420-1424.

Moses, H. & Daubek, H.G. (1961). Errors in wind measurements associated with tower-mounted anemometers. Bull. Am. met. Soc., **42**, 190-194.

Neal, S.B.H.C. (1975). The development of the thin-film naphthalene mass-transfer analogue technique for the direct measurement of heat-transfer coefficients. Int. J. Heat Mass Transfer, **18**, 559-567.

Oliver, H.R. (1973). Smoke trails in a pine forest. Weather, Lond., August 1973, 345-347.

Oliver, H.R. (1975). Ventilation in a forest. Agric. Meteorol., **14**, 347-355.

Ower, E. & Pankhurst, R.C. (1966). The Measurement of Air Flow, fourth edition. Oxford: Pergamon.

Pankhurst, R.C. & Hobler, D.W. (1952). Wind Tunnel Technique. London: Pitman.

Patterson, J. (1926). The cup anemometer. Trans. R. Soc. Can., **20**, 1-56.

Paulson, C.A. (1970). The mathematical representation of wind speed and temperature profiles in the unstable atmosphere surface layer. J. appl. Meteorol., **9**, 857-861.

Perry, A.E. (1982). Hot-wire Anemometry. Oxford: Clarendon Press.

Pope, A. & Harper, J.J. (1966). Low-speed Wind Tunnel Testing. New York: Wiley.

Pratt, R.L. & Bowsher, J.M. (1978). A simple technique for the calibration of hot-wire anemometers at low air velocities. Disa Inf. (Den.), **23**, 33-34.

Raupach, M.R. & Thom, A.S. (1981). Turbulence in and above plant canopies. Ann. Rev. Fluid Mech., **13**, 97-129.

Raupach, M.R., Thom, A.S. & Edward, I. (1980). A wind tunnel study of turbulent flows close to regularly arranged rough surfaces. Boundary-Layer Meteorol., **18**, 373-397.

Redford, T.G., Verma, S.B. & Rosenberg, N.J. (1981). Drag anemometer measurements of turbulence over a vegetated surface. J. appl. Meteorol., **20**, 1222-1230.

Rees, J.P. (1927). A torsion anemometer. J. scient. Instrum., **4**, 311-316.

Rider, N.E. (1960). On the performance of sensitive cup anemometers. Met. Mag., London, **89**, 209-215.

Rutter, N. (1965). Tattering of flags under controlled conditions. Nature, Lond., **205**, 168-169.

Scrase, F.J. & Sheppard, P.A. (1944). The errors of cup anemometers in fluctuating winds. J. scient. Instrum., **21**, 160-168.

Sheppard, P.A. (1940). An improved design of cup anemometer. J. scient. Instrum., **17**, 218.

Sherlock, R.H. & Stout, M.B. (1931). An anemometer for a study of wind gusts. Univ. of Michigan Res. Bull. **20**.

Shuttleworth, W.J., McNeil, D.D. & Moore, C.J. (1982). A switched continuous-wave sonic anemometer for measuring surface heat fluxes. Boundary-Layer Meteorol., **23**, 425-448.

Smith, S.D. (1980). Evaluation of the Mark 8 Thrust Anemometer-thermometer for measurement of boundary-layer turbulence. Boundary-Layer Meteorol., **19**, 273-292.

Sullivan, J.J. (1978). Modern capacitance manometers. Transducer Technology, Vol. 1, MKS Instruments, Burlington.

Tanner, C.B. (1963). Basic Instructions and Measurements for Plant Environment and Micrometeorology. Soils Bull. No. **6**, College of Agriculture, University of Wisconsin.

Tanner, C.B. & Thurtell, G.W. (1970). Sensible heat flux measurements with a yaw sphere and thermometer. Boundary-Layer Meteorol., **1**, 195-200.

Thom, A.S. (1975). Momentum, mass and heat exchange in plant communities. *In* Vegetation and the Atmosphere. Vol. 1 Principles, ed. J.L. Monteith, pp. 57-109. London and New York: Academic Press.

Thompson, N. (1979). Turbulence measurements above a pine forest. Boundary-Layer Meteorol., **16**, 293-310.

Thurtell, G.W., Tanner, C.B. & Wesely, M.L. (1970). Three dimensional pressure-sphere anemometer system. J. appl. Meteorol., **9**, 379-385.

Tomback, I.H. (1973). An evaluation of the heat pulse anemometer for velocity measurement in inhomogeneous turbulent flow. Rev. scient. Instrum., **44**, 141-148.

Unwin, D.M. (1980). Microclimate Measurement for Ecologists. London and New York: Academic Press.

Wucknitz, J. (1977). Disturbance of wind profile measurements by a slim mast. Boundary-Layer Meteorol., **11**, 155-169.

Yaglom, A.M. (1977). Comments on wind and temperature flux profile relationships. Boundary-Layer Meteorol., **11**, 89-102.

Chapter 7: The measurement and control of air and gas flow rates for the determination of gaseous exchanges of living organisms

S. P. Long and C. R. Ireland
Department of Biology, University of Essex,
Colchester, CO4 3SQ, England.

Introduction

Flow rate measurement and control in pipes is a fundamental requirement in any system designed to measure gaseous exchanges between living organisms and the flowing atmosphere in an enclosing chamber. For example, in the measurement of carbon dioxide uptake by a photosynthesising leaf in a cuvette, two system configurations may be used (Sestak *et al.*, 1971; Long, 1981). First, an open gas-exchange system in which air is passed over the leaf and the net influx is equal to the product of the difference in carbon dioxide concentration and the flow rate across the leaf. Secondly, a semi-closed or null-balance system, in which air is continually recirculated over the leaf and a constant carbon dioxide concentration is maintained by an input of carbon dioxide into the system equal to the net uptake by the leaf. In both configurations the determination of carbon dioxide uptake is as dependent upon the accuracy with which flow rate is measured as it is upon the accuracy with which the carbon dioxide concentration is measured. These arguments apply equally to the measurement of exchanges of other gases and for other organisms. Surprisingly then measurements of flow rate in pipes has, by comparison to measurement of individual gas concentrations, received little attention from biologists. Sestak *et al.* (1971) dedicated thirty-five pages to the measurement of carbon dioxide concentration, but to flow rate measurement – just sixteen lines.

Flow rate measurement

Flow rate, like most physical parameters, may be determined by a range of techniques and may be expressed in terms of either mass (Q_m), in units of kg s^{-1}, or volume (Q_v), in units of m^3 s^{-1}. Photosynthetic and respiratory fluxes or molecular fluxes of carbon dioxide and oxygen are most meaningfully expressed as mass or molecular fluxes, thus it is more appropriate to measure mass flow rate, since if Q_v is measured it must be converted to Q_m using concurrent and accurate measurements of temperature and pressure. Direct measurement of Q_m is clearly not possible since it is not practicable to collect and weigh gas as it is for flowing liquids (Ower and Pankhurst, 1977). Direct measurement of Q_v is possible however and forms the basis of the calibration techniques considered later, but none of these methods are suitable for incorporation into a gas exchange system. To determine gas-flow rate the experimenter must resort to the measurement of some physical effect arising from the motion of gas in a tube. Three effects have been found by experience to be suitable: (1) consequent mechanical effects, such as the rate of rotation of a rotor mounted in the stream; (2) pressure changes; and (3) the rate of heat transfer from a heated body in the air stream.

Three recent reviews list some twenty methods for the measurement of flow rate of gases in pipes (Ower and Pankhurst, 1977; Hayward, 1979; Brain and Scott, 1982). However, this chapter is limited to the few techniques which have found application or appear to have potential application in the study of the gaseous exchanges of biological systems. The range of volume rates that it would be necessary to measure in biological gas-exchange studies varies from 1 $mm^3\ s^{-1}$, for carbon dioxide being fed into a semi-closed system for the measurement of photosynthesis by small leaves, to 10 $dm^3\ s^{-1}$, for air in open systems used for the measurement of carbon dioxide exchange by whole plants, swards or animals in enclosures.

Mechanical methods

Variable-area flow meters. The most widely used instrument for flow rate measurements in biology has probably been the variable-area flow meter, but this is now being supplanted in many applications by thermal mass flow meters. However, the low cost, simplicity and the visual indication of flow rate suggest that variable-area flow meters will continue to be used in teaching and in research, at least as secondary flow rate indicators.

Variable-area flow meters, developed from the the ball-in-tube flow meter of Ewing (1924), consist of a transparent graduated tube with a slightly tapered bore, in which the diameter decreases downwards; and the gas flow to be measured passes upwards. A float of diameter slightly smaller than the minimum bore of the tube is forced by the flow rate of gas up the tube to the point where its weight is balanced by the force of gas flowing past it (Fig. 7.1). In a constant volume flow the ball shape is said to be inherently liable to sudden fluctuations in position within the tube or may 'chatter' against the side of the tube (Ower and Pankhurst, 1977). One solution to this problem is the use of a conical float with angled grooves which causes rotation around the vertical axis, giving the float central stability (Ower and Pankhurst, 1977; Brain and Scott, 1982). This type of float is used in several commercial instruments [1,2]. Variable-area flow meters are available from a range of manufacturers for volume flow rate measurements in the range of 30 $mm^3\ s^{-1}$ to about 50 $dm^3\ s^{-1}$, and the standard range of any one meter is usually 10:1 ([1,2,3,4]; Hayward, 1979). The potential for accuracy depends on tube length, float shape and precision of the glass tube, but typical accuracies may be as good as $\pm 2\%$, although greater accuracy of up to $\pm 0.5\%$ may be obtained with very long precision tubes (Hayward,1979). Instruments are calibrated by the manufacturer for a given gas, at a specified pressure and temperature. These calibrations will only change if deposits of dirt are allowed to form on the tube or float, or if the tube or float become damaged or corroded. The calibration will not apply for other gases or even the same gas at different temperatures and pressures. Correction depends on the density of the gas (ρ) and the resistance coefficient of the float (C), both of which may vary with different gases and flow rates. Correction of calibration depends on the use of manufacturers correction graphs or on recalibration for the experimental conditions. However, C for conical floats used in some instruments is nearly constant, probably because their resistance is due

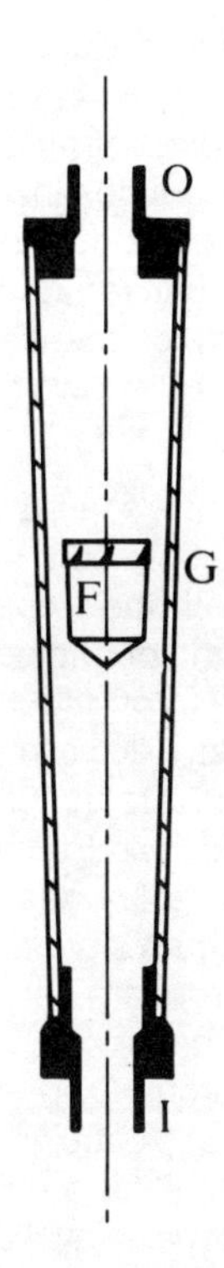

Figure 7.1 Cross-section through a simple area flowmeter, consisting of a tapering glass tube (*G*) and a conical float (*F*) with metal or plastic inlet (*I*) and outlet (*O*) pipe connectors.

largely to the sharp edges of the upper disk rather than surface friction (Ower and Pankhurst, 1977). If *C* is constant then calculation of actual flow rate ($Q_{v,2}$) from that indicated ($Q_{v,1}$) may be obtained simply from the ratios of the densities of the experimental (ρ_2) and calibration (ρ_1) gases:

$$Q_{v,2}=Q_{v,1}\rho_2/\rho_1 \qquad 7.1$$

The correction is important even when the experimental gas is the same as that used for calibration but temperature or pressure differ. For example, if a variable-area flow meter was factory calibrated at 15°C and 101.3 kPa, normal sea-level atmospheric pressure fluctuations from 88 to 108 kPa would produce an error of +7.3% to –3.3% in indicated flow rates. In the field, ambient temperature variations of 0 to 40°C would produce errors of +2.7% to –4%, assuming a constant *C* value for the float. Since flow meters in most gas analysis circuits would be inserted immediately upstream of the assimilation chamber, gas in the outlet of the flowmeter would be above the atmospheric pressure. Either the actual outlet pressure should be measured so that the equivalent flow rate at atmospheric pressure can be calculated or the flow meter should be recalibrated *in situ*.

The major disadvantages of these instruments are that they must be mounted perfectly upright, that subjective errors in assessing the float position relative to the tube graduations are difficult to avoid, especially if the float position fluctuates, and that accuracy is inherently low except in the longest tubes (Hayward, 1979). A further practical problem is that the slightest amount of moisture in the tube may cause the float to stick. The design is not well suited to the production of an electrical output, although instruments incorporating differential transformers to provide electrical outputs for the measurement of flow rate down to 2 $cm^3 s^{-1}$ are available commercially (Farmer, 1980).

Turbine meters. These instruments consist of a turbine or rotor enclosed in a tube within the pipeline, such that the whole flow passes through the rotor. Rotors may consist of flat paddle like blades with the axis perpendicular to the direction of flow (Fig. 7.2) or the blades may be curved or angled to form a propeller with the rotor axis parallel to the direction of flow. The rotor, mounted on low friction bearings, is driven by the force of the flowing gas and, as follows from the theory of the vane anemometer, the rotor speed is directly proportional to the product of the volume flow rate and the square root of the gas density (Ower and Pannkhurst, 1977). Although these meters have found little application in laboratory gas-exchange work so far, they do have some important advantages. Instruments are available for measurements of the very high flow rates, up to 20 $m^3 s^{-1}$. Errors are very low, better than $\pm 1\%$, pressure drop at the turbine is small, down to 20 Pa, and the method is well suited for the generation of analogue and digital electrical output (Ower and Pankhurst, 1977; Brain and Scött, 1982; Studman and Compton, 1983). A design for a simple turbine meter measuring flow rates in the range 1.7 to 150 $cm^3 s^{-1}$ is provided by Studman and Compton (1983) and a wide range of instruments are available commercially [5,6]. Their main disadvantage is the reliance on a moving part. A turbine meter cannot maintain its original calibration over a very long period, because this is bound to change gradually

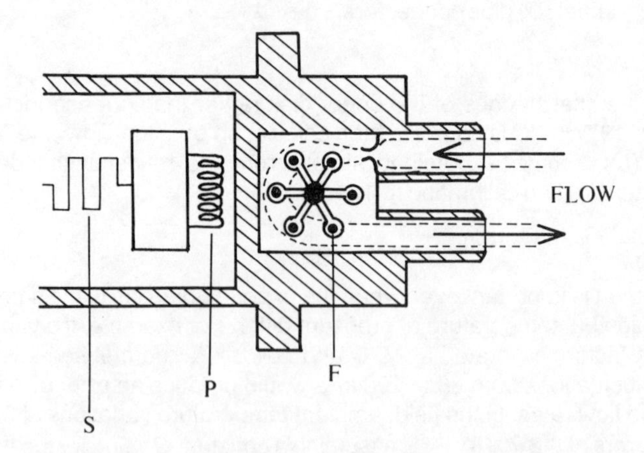

Figure 7.2 Cross-section through a Pelton wheel type turbine flowmeter. The inflow is directed against the flat blades of the rotor and the rotation of ferrite rods (*F*), located at the end of each rotor arm, is detected by a pick-up coil (*P*) producing a digital signal (*S*) which may be totalised and flow rate derived (redrawn from the manufacturer's description [5]).

with wear or loss of lubrication at the bearings, and thus regular recalibration is important. The smaller the size the lower the potential accuracy, since bearing friction will be of relatively greater importance in smaller meters (Hayward, 1979). If the electrical output consists of pulses generated by an electromagnetic or photoelectric pick-up, which detects each passage of the rotor blades (e.g. Fig. 7.2), then care must be taken that electromagnetic radiation generated by power cables, motors, thermostatic equipment etc., does not produce further pulses in the instrument (Hayward, 1979).

Pressure difference methods

The pressure drop that results when a gas passes through a constriction in a pipeline provides another group of methods for measuring flow. Provided that the velocity of flow through the restriction is subsonic and turbulent, the velocity will be proportional to the square root of the pressure drop. Thus, if the pressure drop is measured by an associated instrument the flow may be calculated. Constrictions to produce this pressure drop include orifice plates, standard nozzles, and venturi tubes (Ower and Pankhurst, 1977; Brain and Scott, 1982). Another type of constriction which has been used for the measurement of small flow rates in biology is the capillary tube. Here flow will be predominantly laminar and flow will be more closely proportional to the pressure drop itself than its square root. Pressure drops produced by flow through capillaries of known dimensions may be calculated by Poiseuille's equation or the modified equation of Holmgren (1968). This technique is of particular value in measuring the slow flow rate required for adding carbon dioxide into semi-closed systems for the measurement of photosynthetic carbon dioxide uptake. For example, a flow rate of just 0.26 $mm^3\ s^{-1}$ across a 0.3 mm diameter capillary of 100 mm length would produce a pressure drop of 2 kPa, a drop which could be measured accurately with a simple manometer. Although single capillaries will only be suited to the measurement of low flow rates, honeycomb type arrangements of several capillaries conducting the gas flow in parallel may be used to provide high capacity capillary flow meters (Hayward, 1979).

Pressure measurement. A device for the measurement of the pressure drop across the flow constriction is an essential part of the pressure difference flow meter. "U-tube" manometers, reviewed in detail by Sestak *et al.* (1971), provide a simple method for the measurement of this pressure drop. However, the method is not particularly suited to automated reading and conventional "U-tubes" lack sensitivity for very small pressure drops.

An alternative is the use of small differential pressure transducers. Fig. 7.3 illustrates the construction of such an instrument. When a pressure difference is exerted across the instrument the two diaphragms will be differentially deflected and thereby establish a differential capacitance proportional to the pressure difference. Commercially available instruments can measure pressure differences down to 2 Pa [7]. The major potential source of error is due to temperature induced changes in capacitance characteristics which may introduce errors of up to 0.05% K^{-1}. This error may be reduced by use of compensatory electronics or by precise temperature control of the transducer (Sullivan, 1978). Non-linearity of response to pressure, a significant problem in older pressure transducers, has now been reduced to as little as ±0.05% in some modern instruments (Sullivan, 1978). Thus, capillary flow meters incorporating differential pressure transducers could provide an extremely sensitive and accurate method for measuring flow rates.

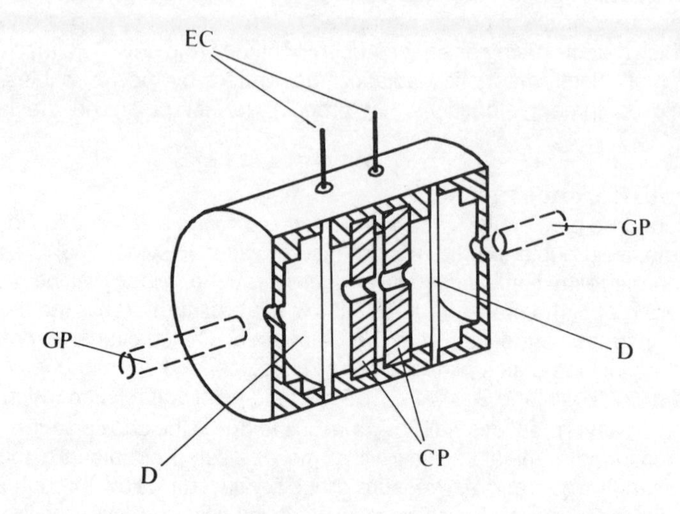

Figure 7.3 Cross-section through a differential pressure transducer containing two capacitors each consisting of a diaphragm (*D*) and parallel electrode plate (*CP*). Two gas pipes (*GP*) connect the diaphragms to the points in the pipework between which pressure difference is to be determined. Pressure difference is indicated via electrode connections (*EC*) to an electrical circuit for capacitance measurement (redrawn from Sullivan, 1978).

The main advantages of pressure difference flow metering are that the small volume flow rates required in some biological gas-exchange research, i.e. $<10\ mm^3\ s^{-1}$ may be measured accurately. Further, the instruments include no moving parts and flow rates may be accurately predicted for standard constrictions without recourse to calibration. Their disadvantage is that since the pressure drop is proportional to the square of the flow rate, rather than flow rate itself, the effective range of these instruments is considerably less than for other flow meters, being typically less than 4:1. This does not apply in the special case of capillary flow meters (Hayward, 1979; Brain and Scott, 1982).

Heat dissipation methods

Thermal mass flow meters. Thermal mass flow meters utilise the thermodynamic principle that the heat carried by a gas flow is related to the heat capacity of the gas and the mass of gas flowing. They consist of a sensor tube which carries the flow and is precisely heated such that the temperature distribution is symmetrical about the mid-point (Thomas, 1911). This may be achieved either by applying heat at the mid-point, as in Fig. 7.4, or by uniformly heating two points equidistant from the mid-point. Two temperature sensors, typically platinum resistance thermometers or thermocouples, are situated one each side of and equidistant from the mid-point. With no gas flow rate the temperature at both sensors will be equal. However, gas flow rate will transfer heat downstream, causing the temperature distribution to become asymmetric and the sensor downstream of the mid-point will then be at a higher temperature. The magnitude of the temperature difference will be a function of the flow rate. These instruments are normally factory calibrated for one gas, but single correction factors, dependent on the molecular structure of the gas and its specific heat capacity, can be used to recalculate the flow rate of other gases; detailed conversion tables are given by Anon. (1981). The first commercial thermal mass flow meters were described by Thomas (1911), more recent developments and calibration techniques were recently reviewed by Widmer *et al.* (1982). As noted in the introduction of this chapter, mass flow rate (Q_m) measurement has fundamental advantages over volume flow rate (Q_v) measurement in the study of gas-exchange by living organisms. However application of the term mass flow meter to these instruments is something of a misnomer. A true mass flow meter measures the mass of gas flowing irrespective of the properties of the gas, angular-momentum meters being the only type relevant in the flow rate ranges concerning the biologist (Orlando and Jennings, 1954; Hayward, 1979). By strict definition, thermal mass flow meters cannot be regarded as true flow meters, since they rely on the thermal properties of the flowing gas and thus meter response will change according to the molecular structure of the gas. However, because heat transfer properties of a gas do not change markedly with changes in its density, they may approximate to mass flow meters for a given group of gasses over a moderate range of temperature and pressure (Hayward, 1979). In practice, performance will suffer less from fluctuations in pressure ($\pm$0.003% kPa^{-1}) than from fluctuations in temperature, which may be $\pm$0.1% K^{-1} between 5 and 43°C (Anon., 1981). This source of error may be reduced by controlling

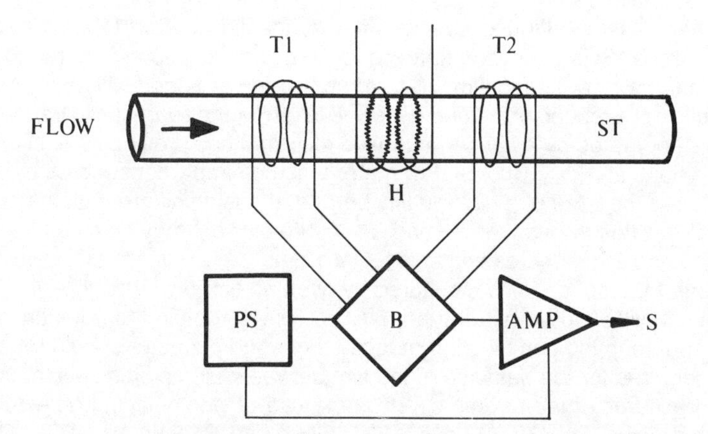

Figure 7.4 A thermal mass flowmeter, consisting of a sensor tube (*ST*) which is heated through a resistance coil (*H*) at the mid-point between two matched platinum resistance thermometers (*T*1 and *T*2). The sensors are connected to a power supply (*PS*), bridge (*B*) and amplifier (*AMP*) to measure, amplify and convert the temperature difference into an electrical signal (*S*) proportional to flow rate (redrawn from Hayward, 1979).

the temperature of the flow meter (Hayward, 1979) or by including a compensating circuit, which measures the temperature of a duplicate reference sensor tube with zero flow rate. This arrangement is now found in at least one group of commercially available meters [9].

Both manufacturers [4,8,9] and independent assessors (Hayward, 1979) suggest accuracies of ±1% of maximum flow rate for thermal mass flow meters, at a given temperature. This has important implications for the application of these instruments in semi-closed systems. For example, if we consider a small system for the measurement of photosynthetic carbon dioxide assimilation by an enclosed leaf area of 10 cm^2 with a maximum photosynthetic rate of 1 mg(CO_2) m^{-2} s^{-1}, then the required range of flows of carbon dioxide into the system would be 0-1 μg s^{-1} or 0-0.51 mm^3 s^{-1} at S.T.P. Since the lowest flow rate ranges of the standard commercially manufactured thermal mass flow meters are 0-20 mm^3 s^{-1} with an accuracy of ±1% of maximum flow, accuracy would be unacceptable at the very low flow rates required. Sufficient accuracy may only be obtained by dilution of the inlet carbon dioxide and allowing a net flow of gas through the system.

Flow meter calibration

It is frequently necessary to recalibrate a flowmeter with different gases or under different operating conditions. It is also advisable to recheck calibrations at regular intervals. For flow rates in the range 10 mm^3 s^{-1} to 10 dm^3 s^{-1} two flow rate measurement techniques have been used as calibration standards: soap-film flow meters and wet gas meters (Ower and Pankhurst, 1977; Hayward, 1979). Widmer *et al.* (1982) reviewed industrial methods for the calibration of thermal mass flow meters.

Soap-film meters

This method, illustrated in Fig. 7.5, was first suggested by Barr (1934). To operate, the rubber bulb (*B*) is squeezed until the level of soap solution in the reservoir rises to the air inlet where a soap film will form across the tube and will be forced up the vertical tube at the speed of the gas. The time taken for the soap-film to travel between two points separated by a known volume is recorded and provides a direct measure of volume flow rate (Q_v). Such a flow meter may be simply constructed by adding a "T" junction to the base of a high-quality burette. By using different diameter tubes a range of flow rates from about 10 mm^3 s^{-1} to 100 cm^3 s^{-1} may be measured with an accuracy of ±0.25%, decreasing to ±1% at 1 dm^3 s^{-1} (Levy, 1961). However, if the movement of the soap-film is monitored by eye and timed with a stop-watch, timing precision is unlikely to be better than ±0.1 s. Thus to obtain the potential accuracy of this instrument, the combination of tube length and diameter should be such that the time required for the passage of the film between fixed points should exceed 20 s. Timing accuracy may be improved by the use of photoelectric detectors which trigger an electric timer ([10]; Levy, 1961; Ower and Pankhurst, 1977). The slight curvature of the soap-film moving up the tube may also introduce some ambiguity into assessing the exact position of the film. This may be overcome by replacing the soap-film with a low friction mercury piston, which is the basis of at least one commercial calibrator [4]. Kolk and Moulijn (1978) have shown the use of such a piston in a high precision diameter glass tube with an anti-static coating combined with photoreflective cells for timing the passage of the piston allows the measurement of Q_v from 50 mm^3 s^{-1} to 50 cm^3 s^{-1} with an accuracy and repeatability better than ±0.25%.

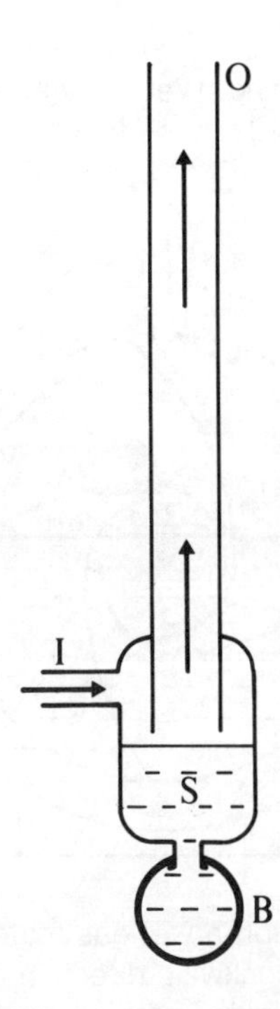

Figure 7.5 Cross-section through a simple soap-film flowmeter consisting of a reservoir of soap solution (*S*) which may be raised and lowered by a rubber bulb (*B*) so that air passing though the inlet (*I*) will form soap films which then pass up a long glass tube bearing volumetric calibrations and the top of which forms the oulet (*O*) to the atmosphere (redrawn from Ower and Pankhurst, 1977).

Wet gas meters

For larger volume flow rates (>10 cm^3 s^{-1}) it is more practicable and accurate to use a gas meter for calibration. Two types of gas meters are used widely in the fuel-gas industry. The dry meter is the basis of the ordinary domestic gas meter and is not sufficiently accurate for use as a standard. The wet gas meter is used as a standard in the gas industry and may provide an accuracy better than $\pm 0.25\%$ (Ower and Pankhurst, 1977).

A sectional diaphragm of a wet gas meter illustrates the operation of this instrument (Fig. 7.6). A rotating drum (*D*), half filled with water, is partially sub-divided into a number of compartments (four, in the illustration) by precisely curved partitions which extend the full length of the drum. Gas is introduced into the centre of the drum from an inlet pipe and fills the compartments. Whilst filling, a compartment is sealed by the water. Filling forces the drum to rotate and when full of gas the curvature of the partition is such that water will seal the compartment inlet and open the outlet. Further rotation will force water back into the compartment displacing gas from the compartment and out through an exhaust outlet at the top of the instrument. The revolutions of the drum may be recorded on an external dial or the speed of rotation converted into an electrical signal for remote reading. Wet gas meters are available commercially for measurement of flow rates from 10 $cm^3 s^{-1}$ to 5 $dm^3 s^{-1}$ [11].

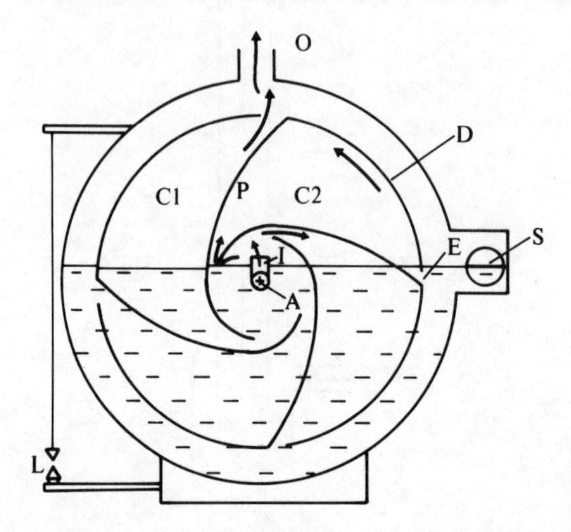

Figure 7.6 Cross-section through a wet gas meter consisting of a drum (*D*) which rotates in an anti-clockwise direction around its axis (*A*) when gas gas flows. The drum is divided into four compartments (*C*) by curved partitions (*P*) each with a gas exit (*E*). Gas enters the compartments by a central inlet (*I*) and escapes through the exit (*E*) when the compartment has rotated above the water level. The arrows indicate the directions of gas flow through the instrument. A plumb-bob level (*L*) and sight-glass (*S*) are provided so that the instrument may be maintained in a perfectly horizontal position and the water level maintained at the correct height (redrawn from a manufacturer's description [11]).

These instruments require previous calibration, which is normally achieved by measuring the rate of water displacement from standard bell-jars (Hayward, 1979). Provided that the instrument is well maintained and is used correctly there is no reason why the calibration should change. The water level must be accurately maintained and the instrument must be perfectly level. Unlike soap-film flow meters a significant pressure drop (>1kPa) is necessary to rotate the drum. It is therefore best to place the instrument to be calibrated downstream of the drum, unless that instrument requires a still greater pressure drop to operate.

Both of the above calibration techniques utilise water. If dry gas is used in calibration, it should be remembered that wetting will increase the volume at 20°C by up to 2.3%. It is therefore advisable to humidify the gas prior to its passage through the flow meters. Neither of the calibration techniques discussed here are suitable for very high flow rates (>5 $dm^3 s^{-1}$). Standard nozzles or venturi tubes operating over a known pressure drop may be used to provide calibration standards at high flow rates (Ower and Pankhurst, 1977).

Flow control

Although the importance of flow control is less easily quantified than flow rate, it is similarly an essential consideration in gas exchange studies. Fluctuation in flow rate will decrease the resolution with which both flow rate and gas concentration difference may be measured, may cause fluctuations in gas composition if different gases are being mixed or humidified, and may introduce both random and systematic errors into flux calculations. Sestak *et al.* (1971) describe membrane flow regulators which are available commercially for control of a wide range of flow rates from 2 $mm^3\ s^{-1}$ to 20 $dm^3\ s^{-1}$ [1,4]. The main disdvantage of this method is the lack of any simple way of controlling flow rate by an electrical input which would allow remote or microprocessor control and adjustment. Since 1971, two further methods of flow control have been adopted on a significant scale in biological research, critical orifices and electronic control.

Critical flow constrictions

A particular problem in mixing gases, either in the supply to gas-exchange systems or for calibration of infra-red gas analysers and other instruments measuring gas concentrations, is that the flow rate of one gas will be influenced by slight changes in downstream pressure produced by small changes in the pipework and connections, or from change in flow rate of a second gas used in generating the mixtures. If a constriction is introduced into the gas stream such that the gas reaches sonic velocity in the constriction, then that velocity and hence volumetric flow rate will be maintained within very close limits (Hayward, 1979). Thus for any given constriction, once the sonic velocity is reached and provided that the critical ratio of upstream to downstream pressure, sufficient to maintain sonic velocity, is exceeded then volumetric flow rate will be insensitive to changes in downstream pressure. However, mass flow rate may be varied simply by altering the upstream pressure and hence density of the gas. Constrictions may be introduced by small nozzles, venturis or orifices (Brain and Reid, 1973; Hayward, 1979). Parkinson and Day (1979) have shown that a range of orifice sizes would provide a fixed range of constant volumetric flow rates provided that a constant upstream pressure is maintained. This principle is utilised in commercially available instruments which generate a wide range of carbon dioxide and water vapour concentrations for the calibration of infra-red gas analysers [12]. These instruments also provide a simple, portable and accurate method for providing known mixtures of carbon dioxide and water vapour at controlled flow rate to gas exchange systems. Although the selection of gas mixtures is manual, remote control could be obtained simply by replacing the manual switches for the selection of orifices with solenoids. The main disadvantage of the method is that for any upstream pressure only a discrete range of flow rates is possible.

Electronic flow rate control

An accurate control of flow rate may be obtained by linking the electrical output of a flow meter to an electronic flow rate control valve, such that the flow rate is continually monitored and automatically adjusted to maintain a preset flow rate. A wide range of electronic flow rate controllers, many patented, are available commercially [4,8,9]. The selection of a suitable valve will depend on the pressure drop, the range of flow rates and response time required.

The two established methods of electrical control are solenoid and servo-driven valves. In solenoid valves, the valve seat is typically connected to the armature and is lifted away from an orifice as increasing current is supplied to the orifice. In servo-driven systems the valve is operated by a small stepping motor. Both of these methods can give good precision (Table 7.1). More recent developments, now commercially available, are thermal expansion valves and piezoelectric crystal valves.

Table 7.1 Electronic gas flow control valves: typical specifications.

Valve type	Max. flow ($cm^3 s^{-1}$)	Max. inlet pressure (MPa)	Response time (ms)
Solenoid	300	0.7	10
Sevo-driven	3000	14.0	15
Piezoelectric crystal	3	0.4	2
Thermal expansion	3000	3.5	8000

Thermal expansion valves consist of a small thin-walled tube with a ball welded to one end which rests in the gas pipe. The tube contains a small resistance heating element and heat-transfer fluid. When a voltage is applied to this element the tube expands and moves the ball which thus controls the flow rate (Anon., 1981). This design has the advantage of having no moving seals, virtually no moving parts (total travel ca. 0.1 mm) and no friction and may thus be expected to be both precise and reliable. However, the device output cannot completely stop the flow rate and thus a solenoid valve must be added to the line if it is necessary to interrupt the flow during experiments. The slow response time of 6-10s (manufacturer's specification [8]) would be a serious limitation in semi-closed systems if the experimental objectives are the study or record of non-steady state changes in gas-exchange, as for example in kinetic studies of the effects of dark-light transitions on carbon dioxide assimilation.

Piezoelectric crystal valves consist of a viton seal cemented to a piezoelectric crystal which has the property of flexing in response to an applied electrical potential. In the resting position the valve is closed but when a voltage is applied the flexing of the crystal opens the valve by an amount proportional to the voltage. The particular advantage of this design is a very fast response time, which according to the manufacturer is 2 ms (Anon., 1979). The lower power requirements of both of these types of controller mean that they may be operated by computer via a digital-analogue converter (DAC) interface, so allowing control through a preprogrammed set of instructions. These instruments also have the advantage of being readily portable and suitable for use at remote field sites.

Conclusions

Thermal mass flow meters combined with electronic flow rate controllers have provided a major advance in the measurement and control of flow rate in gas exchange studies. Flow ranges for these instruments coincide with much of the range required in biology, although pressure difference and turbine flow meters will be more suited and more accurate for the very low flow rates required in small semi-closed systems and the high flow rates of large open systems, respectively. However, thermal mass flow meters must be used with caution particularly under conditions of changing ambient temperature and at low flow rates, where accuracy may be significantly impaired.

Acknowledgements

C. R. Ireland holds a post-doctoral fellowship provided by an Extra-mural Research Award from the British Petroleum Company p.l.c. whose support is gratefully acknowledged. We thank Drs J. Levi, C. F. Mason and K. J. Parkinson for their comments on the draft manuscript. We also thank Linnet Barnes for photographing our drawings and the many people who have provided information on a variety of instruments, in particular Mr G. Edwards of Alexander Wright and Company, Mr A. R. Mitchell of Chell Instruments and Professor P. G. Jarvis.

Manufacturers

1. G.A. Platon Ltd., Wella Road, Basingstoke, Hampshire RG22 4AQ, England.

2. Fisher Controls Ltd. Brenchley House, Week Street, Maidstone ME14 1UQ, England.

3. Fischer and Porter Ltd., Salterbeck Industrial Estate, Workington CA14 5DS, England.

4. Brooks Instruments, Brooksmeter House, Stuart Road, Bredbury, Stockport SK6 2SR, England.

5. Litre Meter Ltd., 50-53 Rabans Close, Rabans Lane Industrial Estate, Aylesbury, Buckinghamshire HP19 3RS, England.

6. Lee-Dickens Ltd. Rushton Road, Desborough, Kettering, Northamptonshire NN14 2QW, England.

7. MKS Instruments, Inc., 34 Third Avenue, Burlington, MA 01803, U.S.A.

8. Tylan Corporation, 23301 SO. Willmington Avenue, Carson, CA 90745, U.S.A.

9. Teledyne Hastings-Raydist, Hampton, VA 23661, U.S.A.

10. Mast Development Co., 2212 East 12th Street, Davenport, IA 52803, U.S.A.

11. Alexander Wright and Co. (Westminster) Ltd., Precision Works, 28 High Street, Tooting, London SW17 0RG, England.

12. Analytical Development Co. Ltd., Pindar Road, Hoddesdon, Hertfordshire EN11 0AQ, England.

References

Anon. (1979). Vacuum-Pressure-Flow Control Systems. Bulletin PFC-11/79, MKS Instruments, Burlington.

Anon. (1981). Mass Flowmeters. Mass Flow Controllers. Tylan Corporation, Carson.

Barr, G. (1934). Two designs of flowmeter, and a method of calibration. J. Scient. Instrum., **11**, 321-324.

Brain, T.J.S. & Reid, J. (1973). Performance of Small Diameter Cylindrical Critical-flow Nozzles. Report 546, National Engineering Laboratory, Glasgow.

Brain, T.J.S. & Scott, R.W.W. (1982). Survey of pipeline flowmeters. J. Phys., E. (Scientific Instruments), **15**, 967-980.

Ewing, J.A. (1924). A ball-and-tube flowmeter. Proc. R. Soc. Edinb., **45**, 308-321.

Farmer, L. (1980). New materials and components. Flow meter. Rev. Scient. Instrum., **51**, 1144-1145.

Hayward, A.T.J. (1979). Flowmeters. A Basic Guide and Source-book for Users. London: Macmillan.

Holmgren, P. (1968). A device to procure air mixtures with accurate carbon dioxide concentrations. Lantbrukshogsk. Ann., **34**, 219-224.

Kolk, J.F.M. & Moulijn, J.A. (1978). An improved apparatus for measuring volumetric flow of gases. J. Phys., E., (Scientific Instruments), **11**, 259-261.

Levy, A. (1961). The accuracy of the bubble meter method for gas flow measurements. J. Scient. Instrum., **41**, 449-453.

Long, S.P. (1981). Measurement of photosynthetic gas exchange. *In* Techniques in Bioproductivity and Photosynthesis, eds. J. Coombs and D.O. Hall, pp.25-36. Oxford: Pergamon.

Orlando, V.A. & Jennings, F.B. (1954). The momentum principle measures true mass flow rate. Trans. Am. Soc. mech. Engrs., **76**, 961-965.

Ower, E. & Pankhurst, R.C. (1977). The Measurement of Air Flow 5th Edn. (In SI/ Metric Units) Oxford: Pergamon.

Parkinson, K.J. & Day, W. (1979). The use of orifices to control the flow rate of gases. J. appl. Ecol., **16**, 623-632.

Sestak, Z., Catsky, J. & Jarvis, P.G. (1971). Plant Photosynthetic Production: Manual of Methods. The Hague: Dr W. Junk.

Studman, C.J. & Compton, S.E. (1983). A low-cost sensitive flowmeter. J. Phys., E. (Scientific Instruments), **16**, 190-192.

Sullivan, J.J. (1978). Modern capacitance manometers. Reproduced from Transducer Technology Vol. 1, MKS Instruments, Burlington.

Thomas, C.C. (1911). The measurement of gases. J. Franklin Inst., **61**, 411-460.

Widmer, A.E., Fehlmann, R. & Rehwald, W. (1982). A calibration system for calorimetric mass flow devices. J. Phys., E. (Scientific Instruments), **15**, 213-220.

Chapter 8: Remote site recording

F. I. Woodward
Department of Botany, University of Cambridge,
Downing Street, Cambridge, CB2 3EA, England.

Introduction

The subject matter of this chapter is the *in situ* recording of environmental measurements at sites which are remote from sources of mains electricity, with instrument power taken from batteries. The emphasis on *in situ* recording implies the exclusion of telemetry.

The available techniques for recording are diverse and no one solution is universally acceptable for field studies. Consequently this review will aim to describe the currently available range of devices for recording variations in climate and plant processes, such as root temperature, leaf wetness and plant extension, which may also be automatically measured with electrical transducers. In general it has been assumed that electrical transducers will be used; they offer the widest range of sensitivities and they may be readily connected to recorders, or data loggers which record electrical signals.

The range of recorders

The available range of recorders may be classified in many ways, for example recording frequency, storage capacity, cost, battery life and even weight, which can be an important consideration when the recorder must be carried to a distant site. A generally applicable classification has been achieved by considering the maximum sampling frequency of the recorder, and the available recording time between field visits to the recorder for changing or reading the appropriate storage medium. The range of sampling frequency is usually determined by an automatic clock mechanism, which initiates a recording at predetermined intervals. Commonly available recorders have been classified in this way in Fig. 8.1. Five major types of recorder are available and in general use, chart recorders, tape recorders, solid-state memories, integrators and printers.

Chart recorders

Chart recorders are standard pieces of equipment in many laboratories, with a wide range of available recording techniques and frequencies. The range of these recorders for field use is much smaller and two general types may be recognised, parallel and serial recording devices. Parallel devices have a recording pen and amplifier, which are dedicated to one channel, so that a three channel recorder (usually the maximum for portable devices), for example, would consist of three pens and amplifiers.

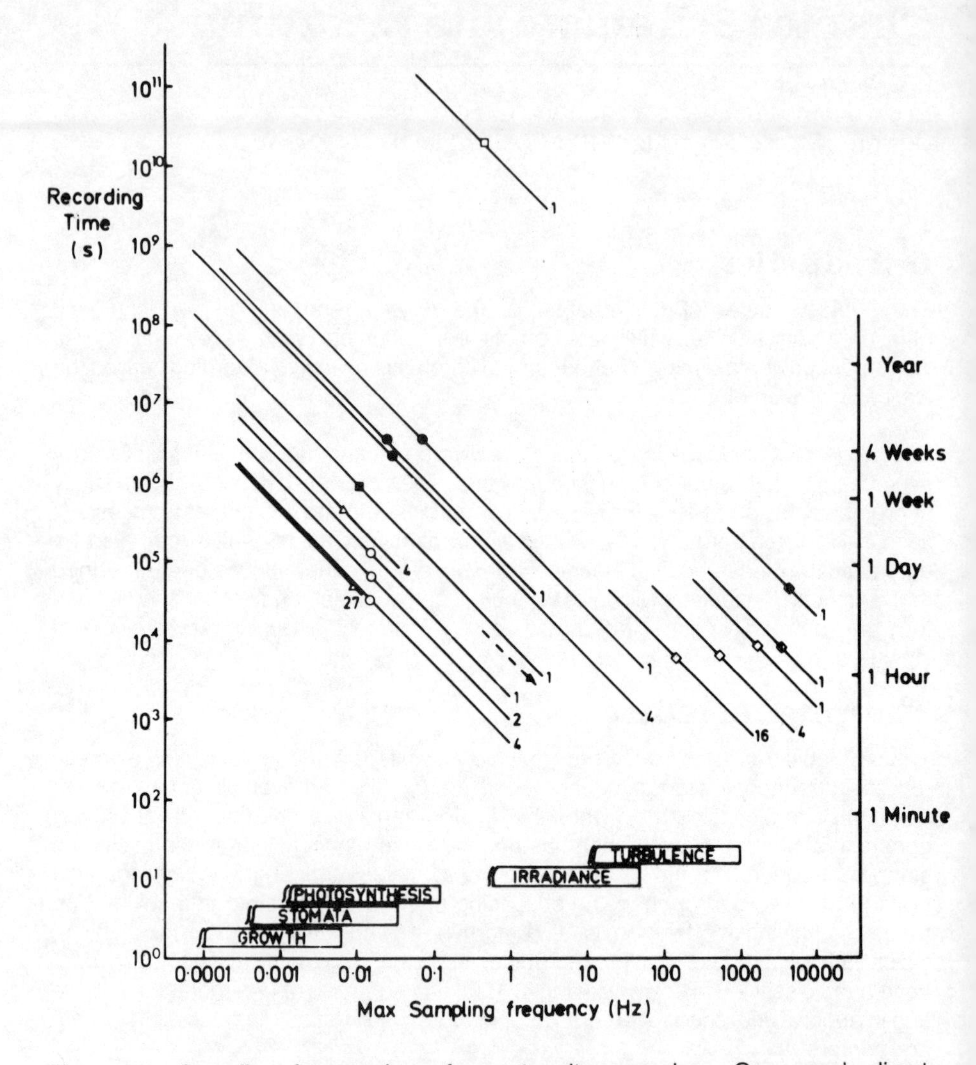

Figure 8.1 Sampling frequencies of remote site recorders. Open reel, direct recording, ◆, FM recording, ◈. and PCM recording, ◇. Cassette or cartridge recorder, ●, serial chart recorder, △, printer, ■, soild-state recorder, ○, and integrator, □. The numbers to the right of the lines indicate the amount of input channels.

The maximum recording frequency of most portable pen recorders is in the range 2 to 5 Hz and would need to operate at its maximum chart speed to resolve rapid changes in an input signal, from the typically rather wide pen trace. The life of the chart in this instance might be in the order of 1 to 2 hours. The chart speed can usually be reduced by 2 to 3 orders of magnitude, increasing chart life. However the current drain of the recorder is usually high and typical lead-acid gel batteries would only last for

1 to 2 days of continuous recording. Intermittent recording is possible with a time switch but pen recorders suffer from large ink blobs on the chart. An automatic pen lift would therefore be required. Rapid changes in the input signal are not readily distinguishable when slow speeds of recording are employed because of the finite thickness of the recorded line and so the effective sampling frequency will decrease. Parallel chart recorders are generally most useful for relatively short periods of recording, up to 1 to 2 days.

Serial chart recorders operate by sequentially scanning each input to the recorder, for a short interval. The process may be continuous or, with a time clock, intermittent. Large numbers of channels may be accommodated in this way, with the record of one channel adjacent to the next and separated by a small gap on the time axis. The serial recorder has a low sampling frequency, up to 0.01 Hz for a 27 channel recorder, and consequently the recorded data are particularly unreliable when the input signal changes rapidly. Commercially available devices may be used in the field for up to 4 weeks, at low sample frequencies.

Tape recorders

The most popular range of remote site recorders rely on the recording of climatic data onto magnetic tape. In mechanical terms the recorders may be classified as either open-reel or cassette and cartridge recorders. Three recording techniques are available; direct, frequency-modulated (FM) and pulse-code modulated (PCM).
Direct mode recorders directly record the input signal but are limited to signal frequencies above about 50 to 100 Hz, below which recorder noise obscures the actual signal. Open-reel recorders can record frequencies up to 2 MHz at a maximum tape speed of 3.04 m s^{-1}. The ratio of the signal to the tape noise voltage (SNR) is about 45:1 at a typical tape speed of 0.76 m s^{-1}, and indicates the ratio between the maximum and minimum detectable signal. The input signal is immediately available when the tape is played back.

The FM recorder is used when signal frequencies close to direct current must be recorded. In this case a fixed frequency, sine-wave carrier signal is modulated by the voltage and frequency components of the input signal. The input signal must be reconstituted by appropriate demodulator circuits on play back. An open-reel recorder can record up to 500 kHz at a tape speed of 3.04 m s^{-1}. The SNR is about 320:1 at a typical recording speed of 0.76 m s^{-1}, and is clearly superior to the direct recording technique.

The direct and FM recorders are devices which are most readily suited to recording rapid changes in an input signal, over rather short recording periods up to about one day. Intermittent use is not usual, probably because of the finite starting time of the tape mechanism (about 2 s) and the possibility of pinch-wheel distortion, if the mechanism remains permanently in the operating position.

The most commonly used and recent development in tape recording, with a wide dynamic range of recording frequencies, is PCM. The input signal to the recorder is converted into an equivalent binary signal. The binary signal is then recorded on the tape at a frequency determined by a synchronising signal. A range of recording codes, for identifying the two binary states and for reconstructing the original synchronising signal, are available (Woodward and Sheehy, 1983) and all have high recording fidelity with errors of 0.0001% or less. The binary signal is usually recorded with a simple error (parity) check providing a simple test for the accuracy of recording. This represents a big improvement over the direct and FM recording techniques. However, as for FM recording, a rather sophisticated data translator is required for reconstituting the original input signal. The tape mechanisms are specially designed to avoid problems of pinch-wheel distortion and the finite starting time may be used as a gap for separating consecutive recordings.

Solid-state memories

One of the problems of remote site recording is that of leaving the recorder unattended and for quite lengthy intervals. It is also true to say that recorders which rely on mechanical mechanisms often break down because of a number of problems which could be readily observed, or even avoided, with more frequent attendance. Examples include, slipping or breakage of motor-drive belts, insufficient lubrication of moving parts, the influence of condensation or freezing and the high currents required for operation, particularly from the stop position. In view of these problems the development and availability of recorders with no moving parts is a welcome sight. The solid-state recorder is one such device. The input signal to these recorders is converted into a digital signal which is then stored in a solid-state random-access-memory (RAM). The maximum sampling frequency of these recorders can be very high and the real limitation is the rather small storage capacity, due more to problems of cost than power requirements. At hourly recording intervals the majority of commercial devices could be left unattended for between 4 (4 channels) and 16 (1 channel) weeks.

The memory will be completely erased if power is withdrawn from the recorder, so current drain, and backup power supplies, must be carefully considered. Non-volatile RAM is now available, with data retention even when the power is withdrawn, however they are more expensive and draw rather high currents.

The records stored in the memory are readily accessible without change to low power displays and so recording fidelity may be easily ensured in the field, but the complete data sets may only be accessed with specialised equipment, and generally in laboratory conditions.

Integrators

Three types may be recognised, the solid-state e.g. Campbell (1974), Buckley (1976) and Woodward and Yaqub (1979); the electromechanical e.g. Macfadyen (1956) and Powell and Heath (1964) and the physical e.g. Ambrose (1980).

The simplest and most limited type is the physical integrator designed by Ambrose. The device measures temperature through its influence on the rate of diffusion of water through a plastic membrane to a Zeolite core. The response time is in the order of a minute but reasonable accuracy (± 0.2 K) is possible. The electrochemical devices integrate small currents which may either e.g. electrolyse silver nitrate to release silver, or cause the movement of mercury menisci in a mercury coulometer. The increase in weight of the silver and the movement of the mercury are integrals of current through the devices. These devices are cheap but fragile and may not readily be used for small currents or voltages.

The most versatile integrators are the solid-state devices. These operate like solid-state memory recorders and commonly convert an input signal to a variable frequency. Instead of storing sampled data in different memory locations, the integrators usually record data continuously by adding counts to one area of memory per channel. The counting frequency is directly proportional to the input signal. The memory may be the display of a low power calculator (Saffell, Campbell and Campbell, 1979), or low power divider circuits (Woodward and Yaqub, 1979).

Printers

Printers have low operating frequencies and so they may be used most successfully as either sampling devices, recording an input signal at predetermined intervals, or as a modification of the solid-state integrators, printing out integrals at selected intervals.

These devices suffer from the problem of drying ink, or if of the electrostatic type, condensation, which must be avoided because the signal to the printing head is shorted and may burn out the printing unit.

Real time checks of recording fidelity are easily achieved but no automatic data translation is possible.

Sampling

It may be seen from Fig. 8.1 that an increase in sampling frequency directly corresponds with a decrease in recording time. This obvious feature reflects a more complex problem in remote site recording and that is deciding, in some objective way, on the appropriate sampling fequency and number of sampling points for a particular project. Within the biological context, it is clear that these properties are related to the biological mechanisms under study. Clearly there is no point in recording at very high frequencies if the only object of interest is plant growth rate, which has a long response time and therefore a low frequency of response.

Shannon (1948) has shown that a minimum of two samples is required for every cycle of a process, if its time course is to be satisfactorily reconstructed at a later date. Cycle, in the biological context may be taken as the inverse of the time constant of a process. So that leaf temperature, with a time constant of 3 s, would have a frequency of 0.3 Hz, and should therefore be sampled at 0.6 Hz. Woodward and Sheehy (1983) have considered objective techniques for designing sampling networks in both space and time and in greater depth than is possible here.

The maximum frequency response of a number of important biological processes is shown in Fig. 8.1. The aim is simply that of deciding on a recorder which is capable of recording the maximum frequency of the process under study.

The lowest frequency response is shown by plant growth and it is clear that chart recorders, printers, integrators and PCM tape recorders are ideally suited for recording. Solid-state recorders might be found to have too limited a recording period. The same range of recorders would be acceptable for comparatve climatic measurements alongside those of photosynthesis and stomatal movements.

Characterisation of irradiance, particularly under a thin plant canopy, and exchange in the turbulent boundary layer, require sampling frequencies which may be in excess of 100 Hz. In this case only tape recorders, probably the open-reel type, with either direct, FM or PCM recording are acceptable, but only for short recording times of a few hours.

The Shannon sampling theorem implies that inadequate sampling of a process will lead to distortion of its true nature, on reconstruction from records. This process, called aliasing, is demonstrated in Fig. 8.2, for temporal changes in photosynthetically active radiation (PAR) over a daylength of 16 hours. The smooth, solid line simulates the change in irradiance on a clear day in late summer. The irregular curve shows the

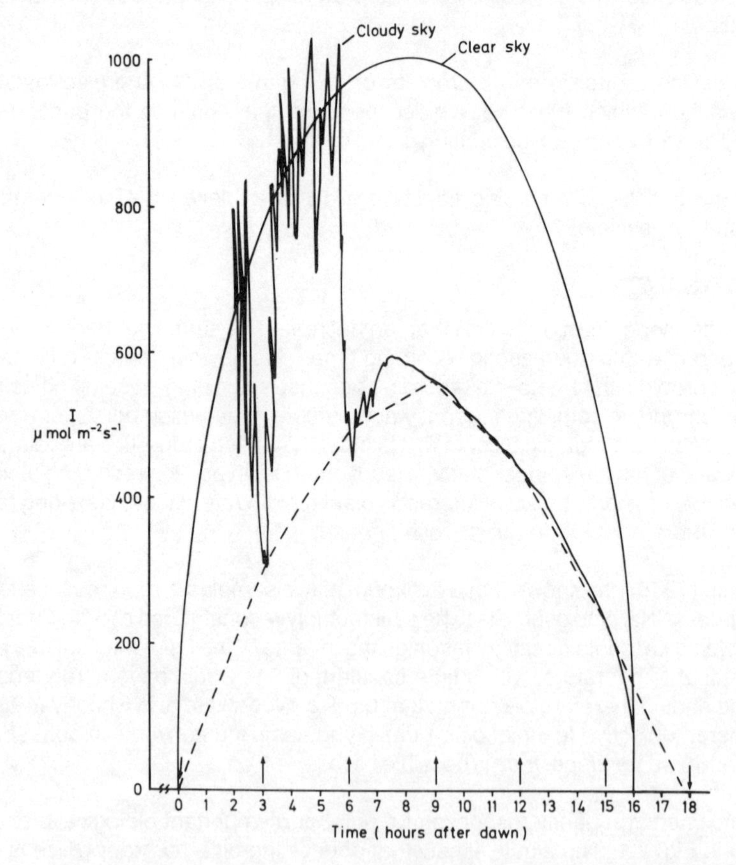

Figure 8.2 Changes in irradiance (I) for a clear sky and a cloudy sky and predicted from 3-hourly records (– – – – – – –). Sampling points (↑).

typical change in PAR which would be expected during a period of cloud, which intermittently obscures the solar beam, between 2 and 6 h after dawn, and leads to a high frequency component in the trace. At 6 hours from dawn the solar beam is completly obscured by cloud for the remainder of the photoperiod. The dashed line shows a reconstruction of the time course of PAR, which would be possible from recordings at 3 hourly intervals starting at dawn. The integral of irradiance is badly underestimated for the first 6 hours, is well simulated for the next 9 hours but overestimates irradiance to dusk.

A true reconstruction of the temporal changes in PAR could be achieved by sampling at intervals of about 5 minutes, for the first 6 hours. Clearly this frequency would be greater than the optimum for the remainder of the day and would lead to an accumulation of superfluous data, although in a way its importance lies in confirming that the later trends in irradiance are smooth with time. In such a situation the recorder should ideally either 1) detect the periods of small changes in environmental conditions and reduce the recording frequency in this period, or 2) apply a smoothing function to the input data. The latter response could be implemented by integration, with occasional recording or printing of these integrals.

The action of integration in this way is very attractive but may again lead to significant distortion of predictions, particularly if the data are to be used to predict a non-linear biological response. This type of problem is shown in Figs. 8.3 and 8.4 for the net carbon dioxide exchange (NCE) of simulated canopies of a species, differing in leaf area index (LAI) and with a transmission coefficient to PAR of 60% per unit LAI. The response curves of NCE to irradiance are shown on Fig. 8.4 for a range of LAI. The response function changes from that with a marked plateau at low LAI, to a curvilinear response at high LAI, with no plateau. These response curves have been used to predict the daily mean NCE, determined by integrating NCE under the clear sky irradiance, shown in Fig. 8.2, with a mean PAR of 767 μmol m^{-2} s^{-1} and under a lower mean irradiance of 323 μmol m^{-2} s^{-1}. The daily integrals have been compared with the predicted mean NCE using the measurement of mean PAR alone, in effect using a daily integral of irradiance to predict NCE. The latter predictor was always greater than the true daily integral of NCE and is shown in Fig. 8.4 for a range of LAI. It is clear that the magnitude of the error changes with both LAI and mean irradiance, casting severe doubt on the use of integrals in this way. If the biological response is known then integrators may be designed with different thresholds, e.g. integrating irradiance over 100 μmol m^{-2} s^{-1} intervals, when non-linearity in NCE is rather small. Such possibilities have been considered by Buckley (1976) and Woodward and Yaqub (1979).

The problems of temporal sampling may be resolved by recourse to continuous measurement of the variable under study for selected periods, and after due consideration to the associated biological response. However no such aids are available for designing networks in space. Spatial sampling is important in two ways, first for the adequate description of local climate and its variation and second as a determinant of the number of sample points and therefore the maximum recording frequency and recording time.

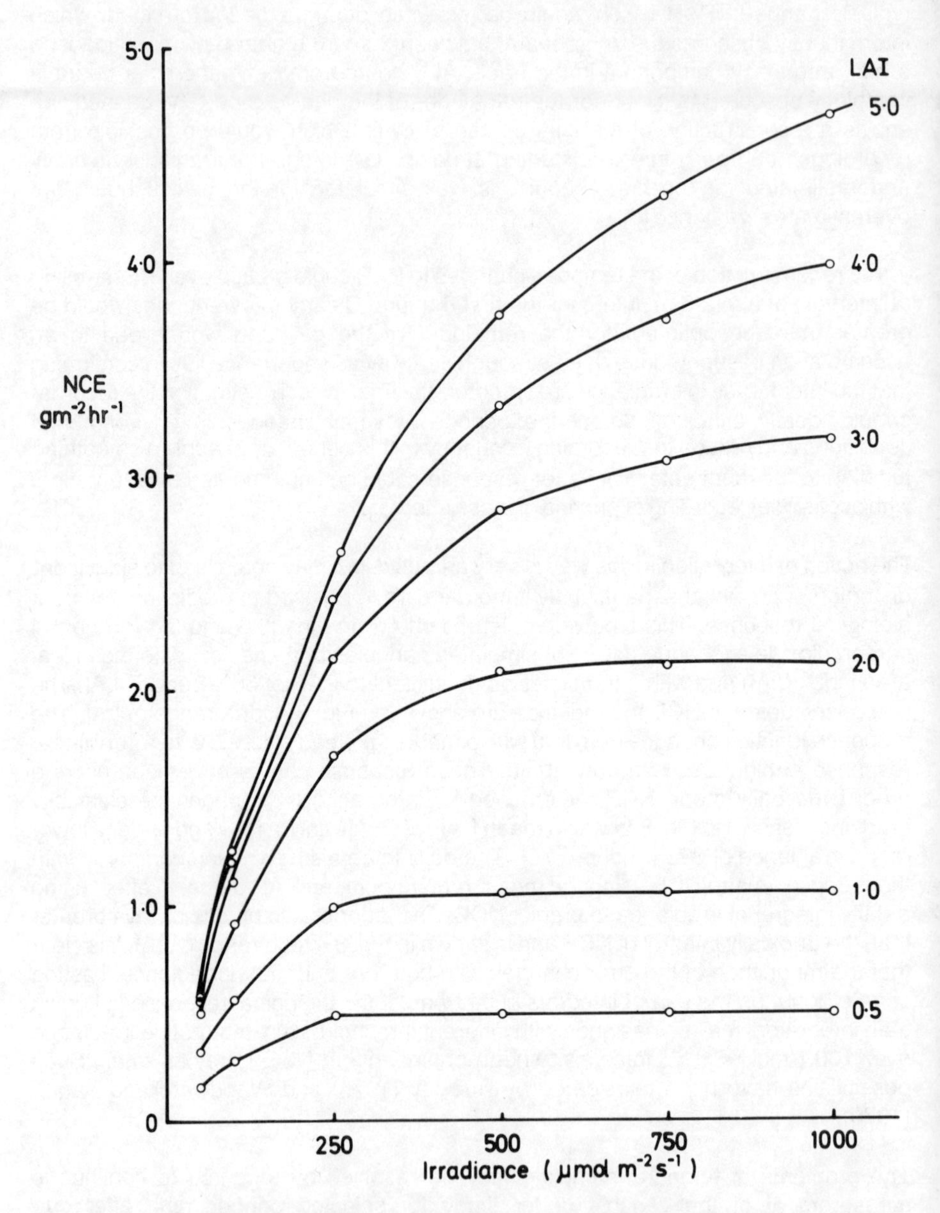

Figure 8.3 Net carbon exchange (NCE) of a canopy differing in leaf area index (LAI), at a range of irradiances.

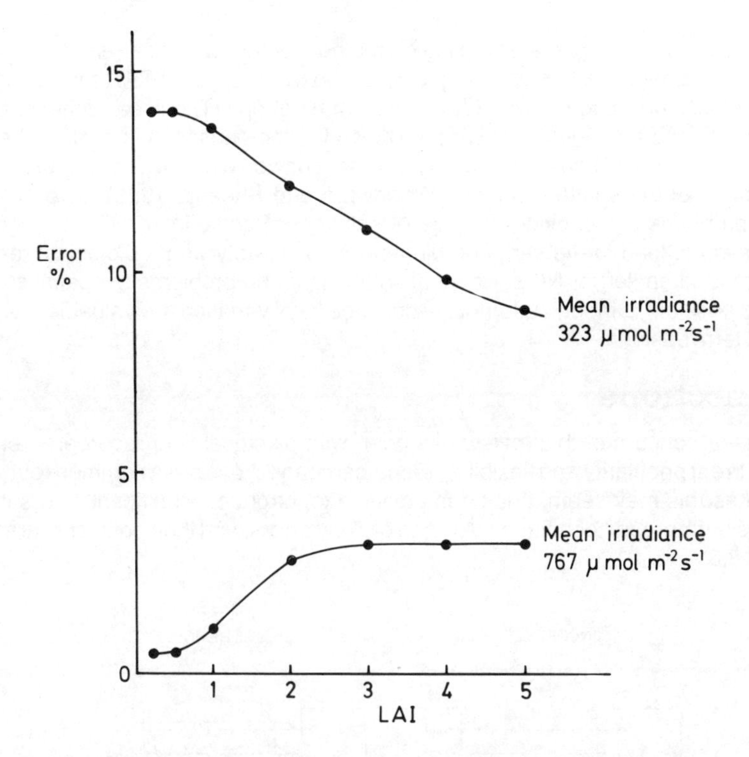

Figure 8.4 Percentage error on the measurement of net carbon exchange predicted from mean daily irradiance, compared with the mean daily integral.

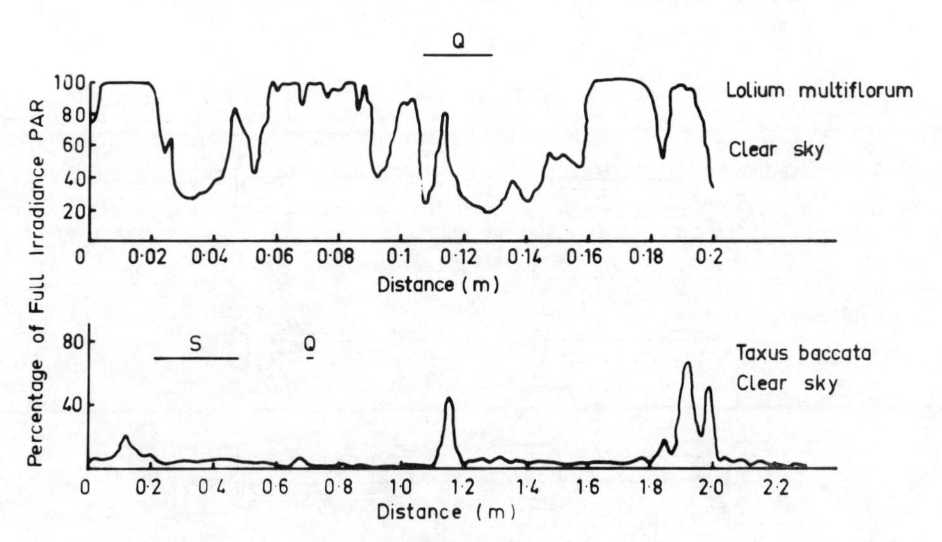

Figure 8.5 Transmission of photosynthetically active radiation along a transect through two canopies. *S*, length of a typical solarimeter and, *Q*, length of a quantum sensor.

A typical problem for spatial sampling on the horizontal scale (the vertical scale must also be considered) is shown in Fig. 8.5, for the penetration of PAR through a grass (*Lolium multiflorum*) and a tree (*Taxus baccata*) canopy. The linear dimensions of a solarimeter and a quantum sensor are shown for comparison. It is clear that either a very large number of sensors, or a very long sensor would be required for an adequate description of the spatial variation. Woodward and Sheehy (1983) have described these problems and provide a range of solutions. Spatial integration by one large sensor would lead to the same problems of non-linearity with a biological response as described earlier for NCE. No ideal solutions to the problems of spatial sampling may be possible, although attempts to describe local variation are valuable, even over a short term basis.

Connections

The major concern of this review has been with electrical field recorders, reflecting their current popularity and flexibility. Electrical connections are a common feature but, simple as this may seem, it is often possible to introduce significant errors into the records by improper techniques. A range of likely errors, and their solutions, are shown on Fig. 8.6.

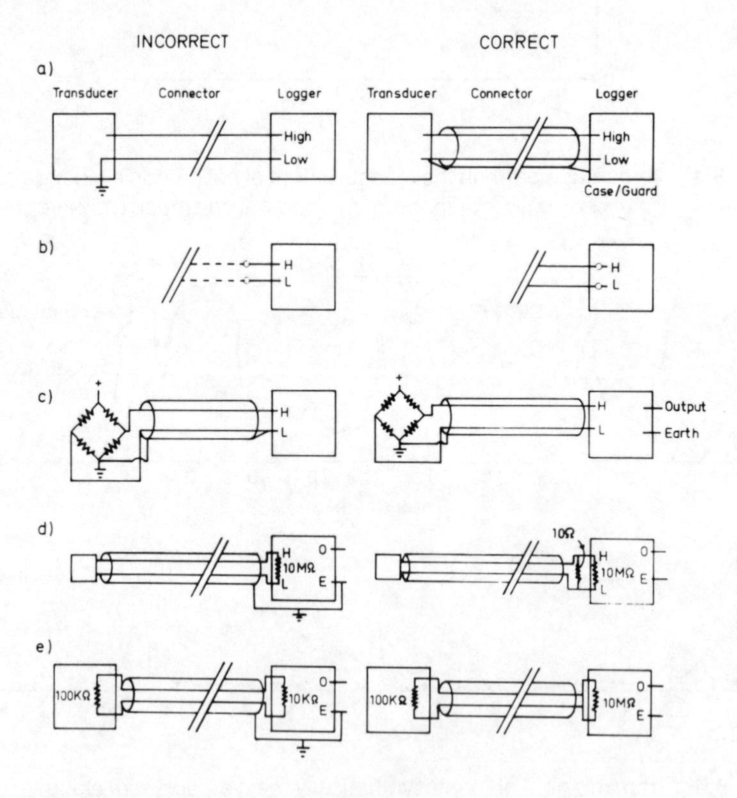

Figure 8.6 Problems in transducer: logger connections.

The electrical signal from a transducer, particularly if it is a small signal, is always liable to degradation through the pick-up of electrical noise. This problem is most severe if the connection between the transducer and the recorder or logger is made by two widely separated conductors (Fig. 8.6a). In this case it is possible that noise will be differentially picked up by the two conductors, leading to the injection of an unknown error into the signal recorded by the logger.

Twisting the wires (twisted-pair) can reduce this problem but a more satisfactory solution, and particularly for small signals (less than about 100 mV) is the use of cable (twisted-pair) within an electrical shield and connection to a floating point on the logger but no signals should be carried on the shield. In spite of this care, noise signals will be introduced into the signal on both conductors, often because of different earth potentials between the transducer and the logger. These potentials arise because of the finite resistance of the shield when both ends are grounded. The mode of connection aims to ensure that these noise voltages are injected equally to both the high and low level inputs to the logger. This noise or error voltage may often be in the range of a few volts. The degree with which this voltage, the common-mode voltage (V_{cm}) is introduced into the recorded signal may be calculated from a knowledge of the common-mode rejection ratio (CMRR) of the logger. This measures the efficiency with which the logger can discriminate against V_{cm} and predicts an unavoidable error voltage (V_e) which will be injected into the signal voltage. The CMRR is defined as:

$$\mathrm{CMRR} = 20\log_{10}(V_{cm}/V_e) \qquad 8.1$$

It is clearly important for loggers to have a high CMRR of at least 100 dB or greater. The rejection of mains interference at 50 Hz, or its harmonics, may be less important at a very remote site than in an industrial environment. It is important to remember that not even a high CMRR can discriminate against a differential injection of common-mode noise into the two inputs to the data logger.

One of the most common and long term problems encountered in remote site recording is that of connections to the logger. The worst case would be when the cable to the logger is a different metal from the connections within the logger (Fig. 8.6b), and with the connector block mounted externally to the logger. Dissimilar metals can act as thermocouples injecting an unknown error voltage into the signal. This injection is only a problem when the temperatures of the high and low connections are different and as a result inject different voltages into the two connections. The simplest solution is to use the same connector, probably copper, throughout. If dissimilar metals are necessary then the connector block should be encased within the logger environment and be of a large mass.

An additional problem with external connectors is that of corrosion of the connecting wires within the connecting block. In some cases electrolytic voltages may occur and in severe cases the connection may be intermittently or continuously open circuit. Clearly the best solution is to house the connector block within the logger and both should be housed in a waterproof housing.

Other problems with connectors in the field are the brittleness of their insulation at sub-zero temperatures, polyurethane, or PTFE being effective to at least –60 °C (Walton, 1982), and the problems of animals which may chew through the cables.

Incorrect operation of the transducer and logger (with no case or guard connection) may be observed on Figs. 8.6c and 8.6e. In the case of the Wheatstone bridge circuit (Fig. 8.6c) the shield connector has shorted one of the arms of the bridge. In this case it is possible to connect the shield to the output earth connection, but remembering that common-mode voltages may be injected because of differences in earth potential. If this problem is considered to be serious, then the screen should not be connected at the logger. Loggers usually have high input impedances which may result in particular operational problems. If the transducer is a photodiode (in the photovoltaic mode) then operation as in Fig. 8.6d will lead to a logarithmic response to irradiance (Woodward and Sheehy, 1983) but a low value, parallel resistance will produce a linear response to irradiance, although at a much reduced input voltage. It is preferable to site the low value resistance at the logger, rather than at the photodiode because any noise voltages will be a smaller component of the larger signal.

A further consideration of the input impedance of loggers may be seen on Fig. 8.6e for two cases with logger impedances of 10 KΩ and 10 MΩ. The input impedance may be considered as a potential divider to the input signal. The equation for determining attenuation by the logger is,

$$1 - I_I/(I_I + I_O) = A \quad 8.2$$

where I_I is the input impedance (Ω), I_O the output impedance of the transducer and A the attenuation coefficient. When the input impedance is 10 KΩ, the attenuation is 91%, which falls to an attenuation of 9% for an input impedance of 10 MΩ. The value of a high impedance is clear in this case. Care must be taken to prevent stray leakage paths between the input to the logger and earth. Loggers with high input impedances should be provided with guard circuits for this purpose. The guard circuits protect the input terminals by intercepting and shorting stray leakage signals to earth.

Recorder characteristics

Six types of field recorder have been described earlier and specific problems which are associated with these devices are shown in Table 8.1 for commercially available devices, where possible.

The column which describes the operating range is self-explanatory. All recorders should be operated in humidities no greater than 95% and should therefore be housed in water-tight enclosures, preferably pressurised, or failing that with a large quantity of desiccant to maintain a low humidity over the full temperature range of operation.

Recorders with rather limited operating temperature ranges should be wrapped in at least two polythene bags and buried in the soil. This serves two purposes; first, large diurnal fluctuations are reduced, therefore reducing the influence of the temperature coefficients of the instruments on the recorded signal, and second, the devices are hidden from view and is about the only security that can be offered at remote sites.

The problems of the input impedance have already been considered. However, the input impedance of the electrochemical analogue integrator is clearly a problem, varying with both temperature (up to 40 ΩK^{-1} at 50 mA) and operating current. The

Table 8.1 Recorder characteristics.

Recorder	Problems	Resolution terminology	Input impedance Ω	Operating range	Typical power requirements
Galvanometric	1. Angular deflection error	FSD	500	−20 to +50°C	
Chart	2. Chart marking		1 M	to 95% RH	75 mW
			(varies with		
Servo	1. Deadband, overshoot		range)	0 to +40°C	3 W
	2. Ink drying			to 90% RH	
FM & Direct	1. Direct recording not less than 50 Hz	FSR	100 K	0 to +40°C	8 W
Tape	2. Non-linearity ±1% of FSR			20 to 80% RH	
	3. Distortion ±2% of signal				
	4. Start/stop time (2s)				
PCM Tape	1. Non-linearity ±0.1% of reading	A/D	10 M	−20 to +50°C	10 W (recording)
	2. Start/stop time (1s)			to 90% RH	3 mW (standby)
	3. High power for tape mechanism				
Integrator:	1. Variations of impedance with	FSD	300-1000	−50 to +70°C	Drawn from transducer
Analogue	current and temperature				
	2. Fragile				
Digital	1. Loss of data with power failure	FSR	1 M	−5 to +40°C	3 mW
				to 95% RH	
Solid-state	1. Loss of data with power failure	FSR	10 M	−10 to +50°C	75 mW
		A/D		to 95% RH	
Printer	1. High power for printing mechanism	FSR	1 M	−5 to +40°C	3 W (printing)
	2. Ink dries	A/D		to 95% RH	2 mW (standby)

simplest solution is the use of a large resistor, typically 10 KΩ, in series with the transducer (Brown, 1973). Although the analogue integrator has clear disadvantages with respect to input impedance it may be operated at very low temperatures. Many recorders may be used at sub-zero temperatures, however long-term use of tape recorders at sub-zero temperatures may result in tape breakage, through brittleness (Walton, 1982).

The power requirements are largely self-explanatory, with the choice of battery being related to the recording period which is required and to the power drain on the logger. Both primary and secondary (rechargeable) batteries are available and all can be used to at least −10 °C. Lithium primary cells may be operated to −40 °C and some of the rechargeable lead-acid batteries may be operated to −65 °C (Walton, 1982). The major consideration in the choice of batteries is the trade-off between the cheaper, one-off use of the primary cell against the more expensive, but repeated use (up to 700 cycles of drainage and charge) for the rechargeable types. Another important feature of battery use is the influence of a declining battery voltage on the operation of the recorder. The influence may be through a reluctance to operate high power requiring circuits e.g. tape or printing mechanisms, or just a general drift in the electronic characteristics of the recorder. The latter effect may be accounted for by using one channel of the recorder for checking zero drift or battery voltage.

It is clearly important to extract the data from solid-state recorders or integrators before the battery voltage falls to the level at which the solid-state memory loses its records.

The resolution of the different recorders may be characterised in three ways, full scale deflection, FSD, full scale range, FSR, and in terms of resolution of an analogue to digital conversion (A/D). All three characteristics must be treated with some care.

FSD is a measure of the displacement of a pen of an analogue chart recorder. If the recorder has an error of ±0.25% of FSD and has a full scale displacement of 1000 mV, then the error of the full scale deflection will be ±2.5 mV. However if the input is only 100 mV, the absolute error will remain at ±2.5 mV and the relative error will be ±2.5% of the reading. The temperature coefficient of the recorder will usually be quoted in terms of the FSD.

FSR is applicable to recorders which convert an analogue voltage to a variable frequency e.g. the digital integrator of Woodward and Yaqub (1979) and in all respects is equivalent to FSD but in terms of the maximum obtainable frequency.

A/D converters are integral sections of digital tape recorders and some solid-state recorders. The resolution and full scale are measured in binary bits. Thus an 8 bit converter has a maximum range of 255, or from 0 to $2^8 - 1$ and a resolution of 1 bit in 256, or 0.4% FSD. So that if the full scale is equivalent to 2.5 V and the minimum is equal to 0 V then 1 bit is equivalent to 0.01 V. Clearly, 2 bits will be equivalent to 0.02 V and there will be rounding errors for voltages between these two values. One obvious problem will be that of matching this limited range and resolution with the incoming voltage. It would not be acceptable to use the full scale voltage of 2.55 V when observing input voltages of about 0.01 V. Increasing amplification of the input signal so that the full range is, for example 0 to 255 mV (an amplification of x10) would prove to be more acceptable, until the input voltage exceeds 255 mV. This problem may be resolved by choosing an auto-ranging A/D converter, which switches between full scale ranges, in response to the input signal. The sampling frequency would then be reduced. An alternative technique would be to choose an A/D converter with greater resolution such as 2^{10} (1 in 1024) or 2^{12} (1 in 4096), but the costs are greater and the sampling frequency may be lower.

The A/D converter of a logger is often classified in terms of the accuracy of a reading (e.g. ±0.1%) and that of the full scale range (e.g. ±0.05%). It is important to realise that these are likely to be the smallest errors and that significantly larger ones are possible. In this case, if the A/D converter has a maximum range of 4096 (a 12 bit or 2^{12} converter), then the error of the range will be ±2 and will be present on all readings. The error in a reading close to full scale will be ±4. So the maximum error could be +6, which is a percentage error of 0.15% of the reading.

At a smaller reading of 1000 the maximum error due to the constant range error will still be +2, while the error of the reading will now be +1, a maximum of +3 or 0.3% of the reading. The error is likely to be somewhat less in practice but this and early discussions of FSD indicate that the maximum possible deflection of the recorder leads to the smallest percentage error.

A number of specific problems may be recognised for the recorders. The charts of chart recorders are sensitive to physical damage and marking, a particular problem for those recorders with pressure-sensitive paper. The ink of ink recorders often dries at an inconvenient time, if not maintained frequently. The same problem is true for ink printers.

The galvanometric recorders are essentially modified analogue meters and the printer of the meter describes an arc over the chart in response to an input signal. This non-linearity may be in the order of ±1 or ±2% of FSD.

The servo or potentiometric recorder operates by comparing the input voltage with the voltage across a potentiometer, provided from a regular power supply. If a difference is sensed then a motor drives the writing pen across the chart and changes the position of the potentiometer, until the voltage difference is nulled. The response frequency (to about 10 Hz), degree of pen overshoot (up to 1% FSD) and the minimum detectable change in input signal or deadband (0.1% FSD) are all features of recorders with this type of response.

Tape recorders in general suffer from the problems of distortion due to fluctuations in the tape drive mechanism and these may be significant for a direct FM recording, particularly when small changes in recorded frequency are detectable as input signals. The nature of PCM recording in a bit sequence, leads to a method of recording with less sensitivity to speed fluctuation and therefore less distortion.

Data translation

The need for field validation of recording integrity has already been stressed for all tape and solid-state recorders, which have no obvious visual check. The other recorders have some form of readout, so that field checks may be easily achieved. A similar type of field validation is also required for data loggers with programming capacity under software control. These recorders are under microprocessor control and the operator may choose a set of operating instructions such as variations in recording frequency for different transducers, with data conversion to standard units and perhaps a variable sampling frequency on a diurnal basis. A check on the latter may be possible by rapid cycling through the diurnal recording cycle. However it is more important to provide a check on transducer performance, so that their correct operation is ensured, at least at the time of each visit to the field. Clearly this check is important for all types of recorder.

Laboratory translation of field data is the only means of analysing lengthy data sets from remote site recorders. The most tedious records are those provided by chart recorders, particularly for devices recording information from more than one channel. Commercial devices are available for manual readout, optical readout and direct computer interface. A number of designs suitable for construction have also been published e.g. Goncz (1974), Rafarel and Brunsdon (1976) and Woodward (1977).

Microcomputers are sufficiently cheap for them to be readily available for laboratory use. Whenever possible, the translation of field records should include a microcomputer and be designed for automatic operation and statistical analysis, e.g. the provision of such features as mean, maximum, minimum and integrals in standard units and even with automatic graph drawing facilities.

Tape recorders may be readily analysed in this way, with no requirement to bring the recorder in from the field. Some solid-state recorders need to be taken from the field to the laboratory for data analysis and this is generally unacceptable because of time wasting and gaps in field records. Intermediate storage, such as onto magnetic tape is possible, but each transfer is likely to increase the possibility of errors, particularly if one transfer is carried out in the hostile field environment. The ideal technique in this instance is the provision of a detachable memory, with its own integral power supply, which may be exchanged for a new memory in the field and then returned to the laboratory for translation.

Printing calculators provide the most satisfactory method for checking field operation but laboratory translation cannot be readily automated and so translation is slow and tedious.

The future

Perhaps the majority of the considerations for the future are tied in with the application of microprocessors to controlling the logger operation. An important facility which should be readily available is a check on the input specification of the logger, e.g. amplifier gain, zero drift, power supply checks and the fidelity of data recording. The first feature, amplifier gain, has often to be taken for granted because no automatic checks are instigated but the feature would be a welcome improvement.

The inclusion of a microprocessor with the potential for a logger with improved flexibility may also prove to be counterproductive, if used carelessly. For example, in many instances it is possible, and often desirable, to minimise data recording in order to extend the recording period. However the method of achieving this requirement must be carefully considered. One way may be through a reduction in the sampling frequency but the problems of aliasing, as described earlier, must be avoided. If the maximum frequency that a sensor can follow is known, its sampling frequency should be twice this frequency. This information must be recorded in the software, for each channel.

Another valid technique for reducing data recording is to record the input voltage and time. A new record is only stored when the input signal changes by more than a preset minimum from the previous stored reading. In this way no more than one record is taken during a period of constant conditions.

The above technique would be of value when complete records of climate are required. If the aim is for a general analysis of local climatic conditions then, provided the appropriate biological response is known, integrals may be obtained, either over the complete environmental range, or within selected ranges over which the observed biological response is close to linear.

Field data reduction by microprocessor with the aim of providing a limited number of records e.g. mean values, may be useful but probably represents an over-indulgent attitude to data recording, undervaluing a sophisticated expensive device. The non-linearity of many biological responses to climate, the use of field data for programming controlled environment experiments and the absence of records for maximum and minimum, may prove to be potentially interesting areas of research for later consideration – but are not possible with the excessive reductionist approach. An improved approach in the case of either an uncertain research plan or a plan with rapid developments, would be a higher recording frequency, with the data being stored onto tape, which may be automatically analysed and reanalysed in the laboratory, on a number of occasions.

If the reduction of data is a clear objective, with no subsequent developments, then recording is best achieved by cheap integrators, recognising the problems of non-linear biological responses but with the benefits of a higher frequency of spatial sampling and realisation that vandalisation and theft are not quite unacceptable problems.

References

Ambrose, W.R. (1980). Monitoring long-term temperature and humidity. Institute for the Conservation of Cultural Material Bulletin, **6**,36-42.

Brown, J.M. (1973). A device for measuring the average temperature of water, soil and air. Ecology, **54**, 1397-1399.

Buckley, D.J. (1976). A micropower, digital temperature integrator. Agric. Meteorol., **16**, 353-358.

Campbell, G.S. (1974). A micropower electronic integrator for meteorological applications. Agric. Meteorol., **13**, 399-404.

Goncz, J.H. (1974). Computer aided digitisation of chart records. J. Phys., E. (Scientific Instruments), **7**, 20-22.

Macfadyen, A. (1956). The use of a temperature integrator in the study of soil temperature. Oikos, **7**, 56-71.

Powell, M.C. & Heath, O.V.S. (1964). A simple and inexpensive integrating photometer. J. exp. Bot., **15**, 189-191.

Rafarel, C.R. & Brunsdon, G.P. (1976). Data collection systems. *In* Methods in Plant Ecology, ed. S.B. Chapman, pp.467-506. Oxford: Blackwell Scientific Publications.

Saffell, R.A., Campbell, G.S. & Campbell, E.C. (1979). An improved micropower counting integrator. Agric. Meteorol., **20**, 393-396.

Shannon, C.E. (1948). A mathematical theory of communication. Bell syst. tech. J., **27**, 379-423, 623-56.

Walton, D.W.H. (1982). Instruments for measuring biological microclimates for terrestrial habitats in polar and high alpine regions: a review. Arctic Alpine Res., **14**, 275-286.

Woodward, F.I. (1977). A device for translating chart recorder information into digital form. J. Phys., E. (Scientific Instruments), **10**, 213-216.

Woodward, F.I. & Sheehy, J.E. (1983). Principles and Measurements in Environmental Biology. London: Butterworths.

Woodward, F.I. & Yaqub , M. (1979). Integrator and sensors for measuring photosynthetically active radiation and temperature in the field. J. appl. Ecol., **16**, 545-552.

Chapter 9: The effective use of microprocessors in a scientific environment

C. Pinches
Department of Electrical Engineering,
University of Leeds, Leeds LS2 9JT, England.

Introduction

For a computer engineer working in a scientific environment, the question "Can I use a microprocessor to solve problem *x*?" has become part of the daily routine. The answer is usually a very simple yes. Unfortunately it is the wrong question, but hard pressed engineers, appreciating the benefits of such a succinct reply, seldom draw attention to this. There are really two key questions which the scientists should be asking. The first being, "Is the use of a microcomputer the best way to solve problem *x*?"; with the supplementary question, "What resources are needed to solve problem *x* using a microcomputer?", being asked if the answer to the former is affirmative.

The microcomputer's ability to solve a wide variety of problems is unquestionable and stems from the extreme flexibility inherent in a device whose detailed operation is controlled by a series of instructions produced after manufacture by a programmer. It is, however, a mistake to assume that such flexibility is always advantageous. One common side effect is that inexperienced practitioners are often drawn into using microprocessors assuming, wrongly, that no matter how complex or ill defined may be the problem, the solution will be obtained quickly and easily, despite a lack of previous experience and a total lack of support facilities. It is worth keeping in mind that, in its raw manufactured state, the microprocessor has the *potential* to solve a large number of problems but the actual *ability* to do nothing at all.

This chapter seeks to establish some guidelines for the use microprocessors by experimental scientists. It is of course impossible to lay down rules which, if followed, will always result in a successful design and if ignored, will damn a design irretrievably. It is, however, possible to identify some of the more common pitfalls which can befall the newcomer to microprocessor-based designs and to outline a number of strategies which, if followed, at least give a better chance of a successful outcome.

System considerations

One of the key issues when developing a new experimental technique is the availability of suitable equipment with which to instrument and control a range of physical variables. In the past, it has been traditional to construct a large portion of the experimental apparatus from fairly basic components, but as experiments become more complex and demanding this approach has become less attractive. Nowadays, many experiments can be performed with fully integrated commercial instruments. The first rule for any experimenter should be to use a commercially available system wherever one is available. Construction of a purpose built system should never be taken lightly.

When a reputable instrument manufacturer markets a product, it consists of far more than a number of components joined together. Its performance will be known and guaranteed, its operation will be well documented and, perhaps most important, facilities for maintenance will exist. Non-engineers often fail to understand the huge differences that exist between a bench prototype of an instrument and the first production version.

The primary aim of any prototype is to verify the correctness of a paper design. Having established the role of the prototype it is a simple step to infer some of its basic properties:

a) It is a short lived entity often consigned to the junk box after a series of measurements have been taken.

b) It is often constructed by the designer.

c) It is almost always operated by the designer.

d) It never needs maintenance.

These properties do not impair the basic function of a prototype but they would be disastrous if they were present in a production instrument. It should, therefore, not come as a surprise to find that when inexperienced personnel construct prototypes as a "cheap" replacement for a commercial instrument, the results are often found wanting. The most common problem is that a total lack of documentation means that the system can only be operated and maintained by the designer. This may the boost the designer's ego but it does nothing for the success of a long term project.

It must be admitted, however, that many experiments cannot be performed using only off-the-shelf products. Bearing in mind the questions raised above, how should the design of such "custom" systems be approached? The key thing to remember is that although an off-the-shelf system may not be available, the performance requirements should not be relaxed without careful consideration. Such issues as the availability of documentation, reliable operation, and satisfactory maintenance arrangements do not become insignificant just because you have chosen to build your own system.

System integration

In constructing a complex system from simpler components, the engineer's task is primarily concerned with linking components together. Each component has a specification and, on the basis of these, the engineer designs and documents interconnections which will turn individual components into functional systems. Nowadays the definition of a component has become somewhat hazy, and many engineers take the view that any subsystem with a well documented specification can

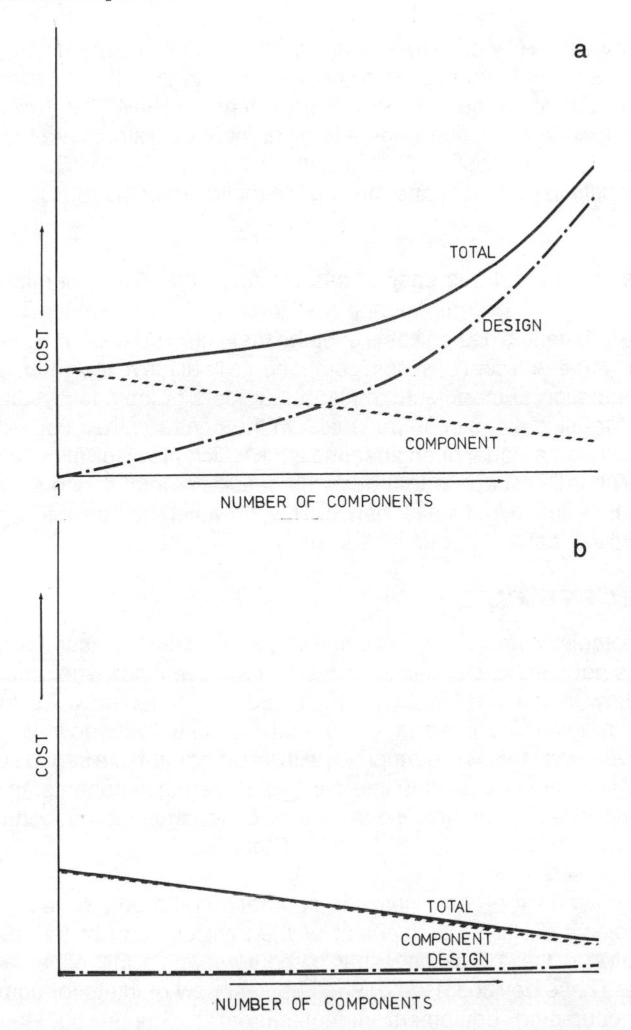

Figure 9.1 The relative design cost, component cost and total cost as related to the number of components for (a) one off system and (b) mass production.

be considered a component. If it is necessary to minimise the engineering effort associated with a project, it is clear that the design should use the minimum number of components since this will, in turn, minimise the interconnection problem and simplify the generation of documentation. When a non-engineer is faced with the design of a one off system, minimisation of the engineering effort required is often the only way to match requirements with available resources. The rule is to use the minimum number of components even though they will be complex and costly.

Of course it is possible to point at successful commercial systems and note that they are manufactured from very basic components. There is, however, a reason for this as Fig. 9.1 shows.

Fig. 9.1a shows the relation between design cost and component cost for a one off system using a variable number of components. Note that the minimum of the total cost curve occurs when the number of components is small. Contrast this with Fig. 9.1b which shows the situation when a large number of systems are to be produced. Here the development cost is divided between all systems produced causing the component costs to dominate and moving the minimum of the total cost curve to the right.

Once this is understood it is easy to see the force that drives commercial system suppliers to use simple components: that force is sales volume. Their aim is to amortise the substantial design costs over the maximum number of systems and then exploit the low material cost of simple components to achieve maximum profit. Where there is no intention of manufacturing large numbers of the final system there is no justification for using simple components. On the contrary, their use will burden the project with a huge amount of engineering work which may well exceed the available resources. The only escape is to reduce the specification and ignore considerations such as documentation and maintenance. This, however, far from solving the problem merely defers the real disaster until the system is in service.

Modular systems

The use of complex components to construct sophisticated measurement and control systems was not practicable until the advent of the computer. The problem was that no matter how much flexibility was designed into the hardware, the variety of interactions required to solve a given problem usually overwhelmed a simple assembly of sub-systems. The computer, with its programmable capabilities, provided the "glue" which could be used to interconnect system components. In this role, the computer manages communications between separate components and takes overall charge of the sequencing and timing of activities.

In the early days of the computer, this approach could only be afforded by large research projects but the development of the minicomputer in the 1960s and the microcomputer in the 1970s have put computers within the price range of most researchers. These devices have opened up vast new markets for computer-based systems encouraging equipment manufacturers to create standard interfaces between their products and computer-based controllers.

Perhaps the best example of such a standard is the IEEE-488 interconnection system, developed by Hewlett Packard in the early 1970s and adopted as an industry standard in 1975.

The striking thing about this interface is the simplicity of the interconnections used in the system. A single, multi-way cable linking all of the instruments is used to convey commands and data between them. In essence, any command which would normally be peformed by operating knobs and switches on a panel can be performed remotely, and any reading which would normally be read off a front panel display by a human operator can be transmitted over the interface for further recording or processing.

Of course merely linking instruments together is not enough to make them execute a complex series of measurements. One of the system components is always a computer and it is the software executing on this machine which gives the system its characteristics. The system designer's major tasks are therefore to choose the correct configuration of instruments and to implement the control program. Notice that there is no need to document either the sub-systems or the means by which they are interconnected since these two areas are adequately covered by manufacturers' documentation and international standard specifications respectively.

The issues regarding the design of the control software are dealt with in a later section but at this stage it is sufficient to note that the control program is only part of the system which must rely on the designer for documentation.

Sub-system design

Despite the rapid emergence of instruments such as those described in the last section, there are still many occasions where the designer cannot fit a commercial device to the problem. Examples may be found at both ends of the complexity spectrum. At the one end, there are projects where researchers are seeking to make "state of the art" measurements. In this case the instrument manufacturers may not be convinced of a general need for the instrument and therefore may not be prepared to commit the substantial sums needed to develop such a device. At the other extreme there are many cases where measurements are needed to a very low precision and the cost of programmable instruments may not be justified. The number of applications which fall into this second category is shrinking rapidly as instrument costs fall. It is a commmon mistake to assume that a commercial instrument is overpriced and that you can do better at less cost. The usual misjudgement lies in comparing the known cost of an instrument which will certainly meet the performance requirements with an optimistic estimate of both the cost and effort required to design a one-off solution. The beginner is likely to assume that the system will work first time, requiring no debugging or systematic testing. Furthermore, "optional extras" such as circuit diagrams and other key documents are not allowed for, with the consequent knock-on effects during the operational life of the system.

It is almost always a mistake to attempt to design a substitute for an instrument available commercially with a view to reducing the total cost of ownership. Even if the commercial device appears to be overspecified for a given application, it may still turn out cheaper at the end of the design phase and will certainly be cheaper to maintain and operate.

Returning to the design of "state of the art" experimental systems, the designer is faced with no alternative but to design a one-off system. There are, however, many ways of approaching this type of design. The approach should follow the same philosophy developed earlier, that is "buy as much as possible – build as little as possible". Fortunately many manufacturers have recognised this problem and have developed products which can be readily incorporated in a custom-built system. These take the form of already assembled functional modules which can be interconnected in a tightly specified manner to form complete systems. The approach is particularly appropriate for computer-based systems since they always require the same basic components, regardless of the application. For example, any computer-based system must have a central processor and some memory. It is possible to buy ready-built modules providing these facilities for almost all microprocessors, obviating the need to resort to the soldering iron.

In the early days of the microprocessor, these modules were very basic in function and each manufacturer produced modules to a different specification, severely limiting the designer's ability to pick and choose from a wide range of options. Nowadays, the situation has improved greatly and the emergence of standards, agreed by groups of manufacturers and in some cases professional bodies, has created a large number of system modules capable of dealing with almost any measurement and control problem.

It is important to understand the implications of a standard and, in some cases, to be aware of their limitations. A good standard should cover both the physical and electrical characteristics of a module. The need for an accurate physical description is obvious; this ensures that modules will all fit into a standard housing. The electrical characteristics are much more subtle since they specify, in considerable detail, the interaction between modules. The bulk of the electrical specification concerns the interface between modules in a system. At the simplest level, signals must be assigned to physical positions on interface connectors. Following this the electrical level specification for each signal must be defined. This merely defines allowable voltage levels for each interconnection; it is important to realise that this is a far cry from ensuring that the system will operate correctly.

Signalling between modules in a computer-based system relies on complex sequences of electrical signals, often requiring tightly specified timing relations. It is in the area of sequence and timing specification that most standards begin to exhibit problems. The major difficulty is that of achieving completeness. As the number of interconnecting signals grows the number of possible sequences and combinations undergoes what is commonly known as a combinational explosion. Taking a simple case with sixteen binary logic signals there are 65536 static combinations and 65536 raised to the power *n* sequences with *n* changes. This simple example takes no account of timing!

The only way in which this problem can be handled is to adopt an extremely rigid formal specification for timing and sequencing, and to impose it regardless of any shortcomings apparent at a later date. It is better to have a standard with limitations than no standard at all. The standards that come out best in this respect are those which have been developed by the larger system manufactures. We have already encountered one highly successful interconnection standard, IEEE 488, developed by Hewlett Packard prior to its acceptance by the IEEE. At the modular computer level several standards exist. One of the most popular standards is the Multibus(tm) module developed by Intel but later adopted by many other companies. The major benefit of this system is the fact that before specification was released to other companies it was refined by internal use at Intel. It was therefore rigidly laid down from the outset leaving other manufacturers, seeking to develop compatible systems, no option but to comply. To understand the significance of this it may be helpful to contrast the Multibus(tm) standard with another popular choice, the IEEE 696 (S100) standard. This latter standard grew out of the products of a small firm which produced one of the first computers based on the microprocessors. The pressure to market the product was such that the initial design was not formally specified prior to its use, neither was it refined by internal use prior to publication. The impact of the product on the market was such that many companies rushed to supply compatible modules.

Very soon, defects in the design became apparent and modifications were proposed and implemented by a number of manufacturers. Although never intended as a standard, the explosive growth of the personal computer based on this system created enormous pressure for the system to be adopted by the IEEE. In the adoption process the standard was extended and modified and finally published as IEEE 696. As it stands, this document has all the characteristics of a workable standard but many manufacturers, particularly those producing equipment before the standard was adopted, produce modules which claim to be compatible but which in practice are not. This chaotic situation is improving with more and more firms complying with the letter of the standard, but it will be several years before true compatibility is achieved.

Software design

As we have seen earlier, the problems of hardware design for one off systems can be minimised by judicious use of ready-made subsystems to perform many common functions. Where software is concerned, however, the situation is not nearly as straightforward. Although modular software products are available and, as we shall see in a following section, can be used to great effect to simplify the design of certain complex systems, the proportion of software which must be created uniquely for a given system is very large. This is simply because it is the software which controls the interaction between hardware subsystems, and it is their interaction rather than their individual function which makes one system different from another.

Many beginners see the task of software development as merely writing a computer program. In reality writing the program, or coding as it is commonly referred to, occupies only a small part of the time taken to develop a working software system. In a key paper "Perspective on Software Engineering", Zelkowitz (1978) produces the

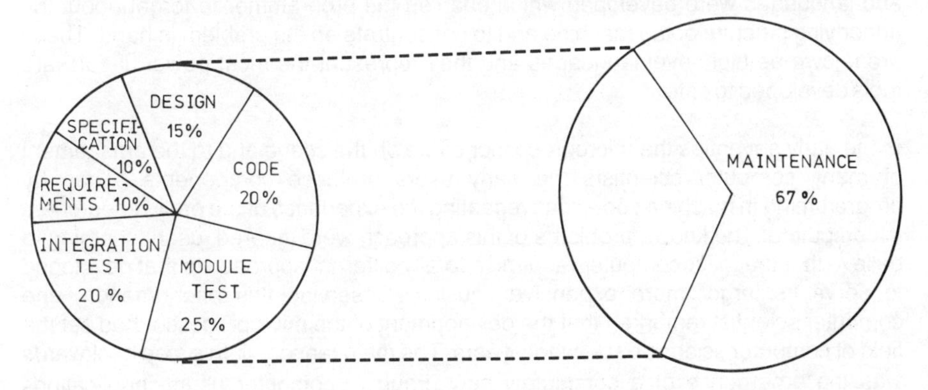

Figure 9.2 The relative effort spent in each phase of the development and maintenance of software. (Redrawn from Zelkowitz, 1978).

two pie charts shown in Fig. 9.2. These results, based on a substantial number of real projects, show that in practice only 20% of the total development effort is spent on coding while some 35% is required for the preliminary stages of requirements analysis, system specification and system design. The other feature, which tends to surprise the uninitiated, is that 45% of total development effort is associated with testing. Perhaps the biggest shock, however, comes when the total effort expended is considered. From Fig. 9.2 it is clear that effort associated with maintenance dominates all other phases of a software project. The proportion shown in the figure is that which is typical for larger projects. Experience has shown that for smaller projects the maintenance effort may drop to 50% of the overall total but seldom drops below this figure. The prime reason for the domination of the maintenance phase of a project is that errors in the original specification propagate through the system phase and are only detected when the system goes into service. Some of these errors may be disguised as "changes in requirements" but their effect is the same.

The reason for this careful examination of the distribution of effort in a software project is that any proposals aimed at aiding the design process must take into account the fraction of the overall task which will be improved. For example, a design aid which halved the time taken to code a program would have minimal effect on the overall project while a design aid which greatly simplified program maintenance would be of great benefit. In the light of this we can examine some of the key design aids available, software tools.

Software tools

The use of software toools to assist in the development of computer programs dates from only a few years after the first digital computer became operational. It became obvious in this short time that in order to exploit even the simplest computer, the human programmer needed assistance. Initially the tools consisted of assemblers, programs which enabled the substitution of mnemonics for the complex numerical codes needed to specify computer instructions and allowed for symbolic representation of computer addresses. It was soon realised, however, that simple translation of an easy to remember code into machine instruction was not enough, and languages were developed which enabled the programmer to forget about the underlying stucture of the machine and to concentrate on the problem in hand. These are known as high level languages and they represent the most powerful software tools developed to date.

In the early seventies the microprocessor came on the scene and to the amazement of many computer scientists the early users of these components began by programming in machine code, thus repeating the experience of the early researchers in computing. The known problems of this approach were ignored, users seeming to believe that the microcomputer required a totally different approach to that developed to serve its larger more expensive cousins. Observing this phenomenon, one computer scientist remarked that the development of the microprocessor had set the field of computer science back twenty years. The main reason for this step backwards was the emergence of a completely new group of computer users, applications specialists who, for the first time, could afford to harness the power of the computer in their particular field of interest. There is no doubt that in the early days this group of enthusiasts were so anxious to use microprocessors that the development of the necessary tools took second place to the development of new application software.

Most manufacturers were prepared to offer simple tools, such as assemblers, but high level languages and other associated tools were unavailable for the first five years of the microprocessor era. One of the main reasons for this delay was that the characteristics of the early microcomputer were far from those of the large machine and the development of suitable tools could not build on earlier experience gained on larger systems. In particular, although a microprocessor may be perfectly capable of executing a program translated from a high level language, it is often a totally unsuitable machine for developing and translating that program. Recognition of this fact led to the so called host-target environment for software development which has done much to increase the availability of software tools. The technique is very simple. It involves developing programs in an environment which provides a wide range of software tools and then, following the translation process, moving the resulting machine code into the system which will support the execution of the program. The machine which provides the development facilities is known as the host while the machine on which the final software will execute is known as the target. The major benefit of this approach is that most software tools have common usage, independent of the microprocessor in use, while the remainder, such as language translators, must clearly be target dependent but actually have many internal components in common. This means that a new processor can be fully supported in a very short timescale using the host-target approach. It is no accident that independent vendors of microprocessor development systems such as Hewlett Packard and Tektronix have opted for the host-target approach while those vendors tied to a particular manufacturer such as Intel and Motorola have developed systems which are tied to their own products.

So far the only software tool given specific mention has been the language translation tool or compiler. A compiler is a computer program which translates from a language that can be understood by the application programmer to a form which can be executed by the target processor. There are many high level languages available to the application programmer each having different properties. The basis on which one should be chosen for a given application is dealt with in the following section. At this point it is worth noting that the compiler does not actually operate alone in the translation process neither is it necessarily monolithic in nature. The application program is first presented to the compiler as input. The translation process is complex and normally performed in several phases. First, the input language must be checked for compliance with rules of grammar and punctuation, then the meaning of each statement is translated into a standard abstract representation and checked for consistency. The final phase of the compiler then converts the abstract representation into a form which takes account of the abilities of the particular target processor. The point to note is that the early phases of the compilation process, those which check the correctness of the input language, are not specific to a particular target processor but are specific to the language in use. In fact the only part of the compiler which is specific to a given target is the final phase which is known as code generation. Many modern development systems share all phases of the compiler except code generation among all the processors supported. Since the development of a code generator represents less than 30% of the total development of a compiler, this sharing allows rapid generation of compilers for new microprocessors.

Although language translation tools such as compilers are extremely important to the application programmer, there are other facilities which must be provided in any reasonable development environment. The most obvious of these is the facility to produce textural material. Computer programs are merely sequences of characters and the ability to create and alter such sequences is central to the task of programming. The tools which provide the required facilities are collectively known as editors. In the early days of computer programming editing was carried out by first punching holes in cards or paper tape and then performing corrections by substituting cards or splicing tape. As anyone who has ever dropped a box of cards will confirm, this was not the best way to create computer programs and nowadays these techniques have been totally superseded by the use of filing systems based on magnetic storage. Early editing software mimicked the operations previously performed on punch cards in that files were altered on a line by line basis. Modern editors, based on word processing techniques, enable the user to treat a video display as a part of a text file and to manipulate text with immediate confirmation of any alteration. The effect of a good editor on programmer productivity is hard to underestimate. It is undoubtedly the most used software tool, being concerned not only with the creation of programs but also the creation of documentation.

The last group of software tools to be described in this overview are those which are concerned with the management of files. The use of magnetic storage techniques to replace boxes of cards and reels of tape depends heavily on the existence of sophisticated software tools which divide the massive storage capacity of modern disk stores into areas storing relatively small amounts of coded data. These smaller areas are referred to as files and they may be used to store text or information such as that produced by a compiler which is only intelligible to another program. By far the most important feature of a file management system is security against data loss. There are many instances where the loss of a critical file could result in the loss of several man months or even man years of effort. It is the presence of copies which is the main protection against data loss in a file system. On small systems the responsibility for taking copies is left to the application programmer while on larger systems this backup is transparent to the average user, being performed by the system operation staff. No file management system is secure in the absence of file copies since no software technique can guard against hardware failure. Well designed systems should have the capability of regenerating a complete file system, following such a failure, using the backup copies.

Again it is the beginner using a small system who is most at risk. Some programmers fail to take backup copies on a regular basis, trusting that the hardware will never fail. Others take copies but do not check their validity. The use of floppy disks in non-maintained systems provides a good example of the type of disaster which can occur. Such disk drives are known to drift out of alignment over a period of time and, if not checked regularly, it is possible to record data so that it is outside the designer's specification. The user may well be unaware of this problem since a misaligned disk drive will often read data created on the same disk drive with no problem. One day the disk drive fails and has to be replaced by a brand new unit which meets the manufacturers specification. Unfortunately none of the backup disks are readable on the new device and the file system security is destroyed.

Computer languages

In an earlier section it was stated that language translation tools were the most powerful available to a computer programmer. The newcomer to the field is, however, presented with a wide choice of language, each having supporters and equally vociferous opponents. To establish a satisfactory basis for making a choice it is necessary to examine some of the properties of a language which are desirable in designing experimental systems.

First and foremost, a computer language should not merely be considered as an aid to programming coding. It has already been established that the coding process forms only a small part of the total effort expended on a project. To achieve a significant reduction in this effort a computer language should, in addition to assisting coding, simplify the design, testing and above all maintenance phases of a project.

The second major requirement is that of flexibility. Learning to use a new computer language with any degree of effectiveness takes a significant amount of time. It is therefore desirable to avoid having to learn a new language whenever a new project is proposed. This can only be done by choosing a highly flexible language at the outset and avoiding special purpose languages which may be able to cope with one class of problem and not another. Incidentally, this relearning problem applies to all software tools, not just language translators and it is another argument in favour of the host-target approach discussed earlier.

The final requirement is that of availability. Knowing that you should be programming in the new miracle language is one thing, but having access to development facilities which support it may be quite a different matter.

It is dangerous to place this last requirement ahead of the other two at a preliminary phase of a project. It is a fact that the creation of the correct support environment can have a dramatic effect on the outcome of a project and furthermore it is the only the improvement which can be obtained by the simple expenditure of money. In particular, beware the substitution of extra manpower for good support facilities, particularly if the project is already running late. Brooks (1975), in his excellent essay "The Mythical Man Month" likens the addition of such manpower to the use of gasolene as a fire extinguishing agent! The problem is that the writing of software, particularly if the problem in hand involves complex interactions, is a difficult task to divide between programmers. It is much better to expend effort on obtaining a support environment which enables the programming to be undertaken by one good programmer.

The first requirement outlined above may, at first sight, seem difficult to achieve by merely choosing the correct language, embracing, as it does, nearly all aspects of the total project. Over the past ten years the development of two techniques, structured programming and top down design, have produced dramatic improvements in the management of computer-based projects and to capitalise on these developments it is important to choose the correct language. Developments in structured programming date from 1965 when it was suggested by Dijkstra, that the quality of computer programmers was inversely proportional to the number of "GO TO" statements contained in their programs. Although this may appear to be a trivial statement, it recognises one of the worst problems in the development of complex programs; that the unconstrained transfer of control from one point in a program to another results in programs which are difficult to produce, test and maintain. Merely eliminating this type of instruction from a computer language does not provide a solution since many languages rely so heavily on this construct that, without it, they would be useless. The positive side of structured programming is concerned with the development of program structures which allow useful computation to be performed without the necessity for these unconstrained transfers. These are, of course, functions of a computer language and hence the link between structured programming and language choice.

At the present time Pascal is the most widely used example of a language, available for microprocessors, which encourages the use of structured programming. Others include Algol, C and Ada. Unfortunately the most widely used microprocessor language, Basic, was not designed with a view to supporting structured programming although many variants have attempted to "tack on" some of the required facilities. The wide use of Basic stems from the ease with which it can be used to solve simple problems and its widespread availability. It is important to note that the ease with which Basic can be used to solve simple problems does not scale up to large, complex problems. It is a good example of a language which aids the coding process, but has little positive impact on other aspects of a project.

The major benefit of structured programming comes in the testing and maintenance phases of a project. Programs are easier to follow and tracing through the actions of suspect procedures is greatly simplified when the investigator knows that control cannot suddenly be transferred to another unrelated part of the program, never to return.

The technique of top down design extends the programming phase of a project into the specification and design phases, which are normally completed before programming begins, and into the testing phases which can be made to overlap the other development activities. The technique is to begin the task by assuming that a program is being written for a hypothetical computer with a very powerful instruction set. Yourdon (1975), in his book "Techniques of Program Structure and Design", describes an overall task called **GLOP** and postulates that the computer in question possesses an instruction which performs the entire task. Thus to solve the problem, the programmer merely writes the one word program,

GLOP

Unfortunately the hypothetical machine does not exist so the next assumption is that another machine does exist which can perform **GLOP** in three stages, thus a complete program may be written,

```
INITIALISEGLOP
PERFORMGLOPCOMPUTATIONS
TERMINATEGLOP
```

Again the hypothetical machine doesn't exist and the process is repeated for the three subdivisions, treating each as a separate problem. Eventually, by subdividing over and over again the point is reached where the machine required is no longer hypothetical but the one that is available. At this point the problem is solved.

The real benefit of top down design is gained not by merely thinking this way but actually writing the program this way. To follow Yourdon's example your first Pascal program should be,

```
PROGRAM GLOP;

BEGIN
END.
```

This may not appear to be a significant step towards an ultimate solution but it is a correct program and will run – it merely does nothing! Following the example to the next stage the Pascal program would be,

```
PROGRAM GLOP;

PROCEDURE INITIALISEGLOP;
          BEGIN
          END;
PROCEDURE PERFORMGLOPCOMPUTATIONS;
          BEGIN
          END;
PROCEDURE TERMINATEGLOP;
          BEGIN
          END;
BEGIN
          INITIALISEGLOP;
          PERFORMGLOPCOMPUTATIONS;
          TERMINATEGLOP
END.
```

Again the program does nothing but a structure is beginning to emerge. The basis is simple. If you know that something needs doing but you don't yet know what, introduce a procedure with a meaningful name that does nothing, taking care to define and use it correctly so that the relation with other procedures is correct. When you have completed this task for a problem treat each procedure as a problem in its own right and reapply the above rule. The point about actually writing a program this way rather than designing it this way is that at each stage you have a correct program which is evolving to meet the final specification.

At the other end of the design process concerned with testing, top down design can be used with great benefit. Each development in the top down process results in a correct program. The only features missing are the detailed operations performed by the low level procedures which are not yet fully implemented. Running these partial programs may seem pointless but this technique is, in practice, very powerful and forms the basis of top down testing. The technique involves the development of the dummy procedures to prove the correctness of the higher levels of the program. After incorporating a procedure which performs no action at all, the programmer may modify the procedure to print a message which indicates that it has been invoked. The resulting printout will verify that the low level procedures are being invoked in the correct order and perhaps at the correct time. Mistakes at the higher levels within a program are much harder to correct as development proceeds and the prime benefit of top down testing is that it tests these levels first, at an early stage in the design process not when the program is complete.

Although top down design is a very powerful technique, experience has shown that beginners resist using it. The main reason is lack of confidence which drives them to build a small part of the system before considering the high level interactions. Usually the part chosen is the one they know most about and the areas where there is the greatest lack of knowledge are left to sort themselves out at some later date. This approach has been likened to someone who begins to design a house by starting to build at one corner, with no plans or drawings. The outcome in engineering terms is often a direct parallel of what you might imagine the house would finally look like!

References

Brooks, F.P. (1975). The Mythical Man-Month, New York: Addison Wesley.

Dijkstra, W. (1965). Programming considered as a human activity. Proceedinngs of IFIP Congress, 1965. Washington DC: Spartan Books.

Yourdon, E. (1975). Techniques of Program Structure and Design, Prentice Hall.

Zelkowitz, M.V. (1978). Perspectives on Software Engineering, Comput. Surv., **10**, no. 2.

Chapter 10: Porometry

K. J. Parkinson
Rothamsted Experimental Station, Harpenden,
Hertfordshire, AL5 2JQ, England.

Introduction

Porometry literally means the measurement of holes (*poros*, GK). However, from the earliest use of the name porometer by Darwin and Pertz (1911) it has referred to the measurement of the conductance of the epidermis to the flux of gases and only by inference to the degree of stomatal opening.

Since the reviews of the subject by Slavik (1971) and by Jarvis (1971) (in the intervening period there have also been reviews by Slavik (1974), Kanemasu (1975) and Burrows and Milthorpe (1976)), a considerable number of new porometers have been described, some new designs based on old types, others based on new ideas. Apart from new instruments there have been two major advances during this period. The first was the introduction of the humidity sensor made by Vaisala (see Chapter 4). This measures relative humidity, has a rapid response to humidity change, a small temperature coefficient, reasonable stability and absorbs little water. It has come to be used in most porometers requiring humidity measurements. The second was the realisation that because significant quantities of water vapour are readily adsorbed and desorbed by many materials, great care has to be taken to avoid such materials in porometer construction.

For reference to the various types of porometers, I have kept the names commonly used in the past, even though these are not necessarily descriptive of their mode of operation. The major division is between instruments that measure water vapour fluxes and those that measure mass flow or diffusion of other gases. Judged by the number of papers concerned with instruments and calibration since 1971, the most popular type is the dynamic diffusion porometer (24 papers), but considering only the period from 1977, steady-state porometry has come to the fore (13 compared with 11 papers).

A concluding section of this review will consider the measurement of water vapour loss from animal epidermes.

Materials for construction of porometers

In types of porometers in which water vapour fluxes are measured it is essential that the porometer itself is not contributing to these exchanges. The properties of the materials are defined both by the permeability and water absorption (Table 10.1). The permeability P is given by:

$$P=(F/A)/(dc/dx) \qquad 10.1$$

Table 10.1 Permeabilities and absorption properties of common plastics.

Material	Common and/or trade name	Permeability ($m^2 s^{-1}$)x10^{12} @ 298 K & 0.1 MPa			Water absorption
		Water vapour	Carbon dioxide	Oxygen	($g g^{-1}$)x10^2
Cellulose acetate		1140-10000	1.8–25.1	0.3–4.3	1.8
Cellulose regenerated	Cellophane	1444–33000	0.001–0.003	0.0003–0.01	20
Epoxy resins			0.07–1.06	0.04–1.22	0.2–1.8
Poly(6 amino caproic acid)	Nylon 6	53–4292	0.05–0.12	0.01–0.29	3.5–8.5
Polycarbonate		52	7.9	1.29	0.36
Polychloroprene	Neoprene	1368	19.0	1.4–3.0	1–6
Polyethylene –low density	Polythene	68–192	7.6–50.0	0.8–7.1	0.024
–high density		9–19	1.6–3.3	0.84	negligible
Polyethylene terephthalate	Mylar[1], Melinex[1]	86–200	0.076–0.166	0.004–0.034	0.6
Polyfluorinated ethylene propylene	Teflon FEP	38	1.3–7.6	2.9–4.5	negligble
Polyisobutylene isoprene	Butyl rubbeer	30–152	3.9–10.6	1.0–4.3	1–6
Polyisoprene	Natural rubber	111–2280	100–179	11.2–18.2	1–6
Polymethylmethacrylate	Perspex	1900–2837	50.2	12.4	1–2
Poly(4-methyl pentene-1)	PMP	61.6		20.5	<0.05
Polypropylene		28.4–66.2	1.4–6.9	0.59–1.75	0.01–0.03
Polystyrene		454–1000	5.9–30	0.48–18.2	0.03–0.05
Polytetrafluorethylene	PTFE	16–27	1.1–8.3	0.5–18.0	negligble
Polyvinyl chloride	PVC	114–480	0.8–13.7	0.09–9.0	1
Polyvinylidene chloride	PVDC, Saran[2] Propafilm-C[3]	0.4–76	0.01–0.04	0.004–0.01	0.03
Polyurethane		270–2200	2.1–30.4	0.34–3.65	1
Silicone rubber (dimethyl)		470–33000	456–2432	53–505	0.07

Compiled from the following sources: Kammermeyer (1957), Major and Kammermeyer (1962), Lebovits (1966), Waggoner (1966), Yasuda and Stone (1966), Wooley (1967), Crank and Park (1968), Roff and Scott (1971), Bloom *et al.* (1980) and Plastic Material Guide (1966), I.C.I. Ltd, U.K.

[1]I.C.I. Plastics Ltd., U.K.,
[2]Dow Chemical Co. Ltd.,
[3]Propafilm-C is a PVDC/Polypropylene/PVDC film produced by I.C.I. Plastics Ltd.

where F is the flux ($m^3 s^{-1}$) through area A (m^2) in response to a concentration gradient dc/dx ($m^3 m^{-3} m^{-1}$) then P is in $m^2 s^{-1}$. The permeability of plastic depends on the degree of polymerisation, orientation of the molecules, cross-linking and the presence of plasticisers and this explains the range of values in the table. The water absorption is measured as the fractional weight increase following immersion in water for 24 h at 298 K. When the permeability is effectively zero (Table 10.2 after Shepherd, 1973; Dixon and Grace, 1982) then water is adsorbed on the surface only.

Unless the porometer system uses long pipes or thin plastic films only ab/adsorption and not permeability need be considered. Glass is then the best choice and plated

Table 10.2 Surface adsorption by materials after 24 hours at 293 K above (a) water (Shepherd, 1973), (b) saturated sodium sulphate solution (93% RH) (Dixon and Grace, 1982).

Material	Surface adsorption $g\ m^{-2}$ a	b
Soda glass	0.05	0.01
Copper	0.05	
Stainless steel	0.05	0.09
Nickel plate on brass	0.05	0.06
Chromium plate on brass	0.38	
Clean brass	0.38	0.07
Tarnished brass	1.05	0.11
PMP		0.03
PTFE	1.05	0.05
High density polyethylene		0.08
Cross linked polystyrene		0.56[1]

[1]Probably includes absorption.

brass, stainless steel or duralumin (suggested by our own measurements) are suitable metals. If plastic is required then PMP, PTFE, or high density polyethylene should be used (see Table 10.1 for abbreviations). Dixon and Grace found stainless steel not nearly as good as these plastics, but their adsorption figure ($0.09\ g\ m^{-2}$) is greater than Shepherd's ($0.05\ g\ m^{-2}$) and Gander and Tanner's (1976) ($0.005\ g\ m^{-2}$), probably because as they suggest, their sample had a porous surface. It should also be borne in mind that it is easier to machine metals than plastics to a good surface and rough surfaces encourage adsorption (Dixon and Grace, 1982). For transparent windows, glass, PMP, or polystyrene are available.

Considering both absorption and permeability then the best transparent films are made from PVDC followed by those made from polyethylene terephthalate. If it is necessary to pipe the gas for analysis, then suitable tubing would be made of stainless steel, PTFE, polyethylene or polypropylene. Silicone rubber adhesives are preferable to epoxy-based adhesives as they absorb less water.

Gaseous transfer system and units of measurement

Many analyses of this have been done (Cowan and Milthorpe, 1968; Jarvis, 1971). Briefly considering an amphistomatous leaf (Fig. 10.1) we can write:

$$E=(\chi_l-\chi_a)\{((r_s^{-1}+r_c^{-1})^{-1}+r_a)^{-1}_{upper}+((r_s^{-1}+r_c^{-1})^{-1}+r_a)^{-1}_{lower}\} \qquad 10.2$$

where E is the flux density of water vapour from the leaf, χ_a and χ_l are the water vapour concentrations in the air and leaf respectively. χ_l is assumed to be the concentration of water vapour in equilibrium with free water at leaf temperature. This is not an unreasonable assumption (Farquhar and Raschke, 1978) and any error will result in a small increase in the calculated stomatal resistance (r_s). The cuticular resistance (r_c) and the stomatal resistance are in parallel (this combined resistance is the epidermal resistance, r_l) and are both in series with boundary layer resistance (r_a). The boundary layer stretches from the leaf surface where gas exchange is by diffusion to the region where turbulent transfer predominates (Thom, 1968). Its thickness and therefore its resistance depends upon the measurement conditions and for accuracy, it should be small, constant and known. The cuticular resistance is large when compared with that of open stomata (Cowan and Milthorpe, 1968). The stomatal resistance includes the resistance to diffusion through the stomatal pore and the intercellular spaces and any resistance at the water/air interface.

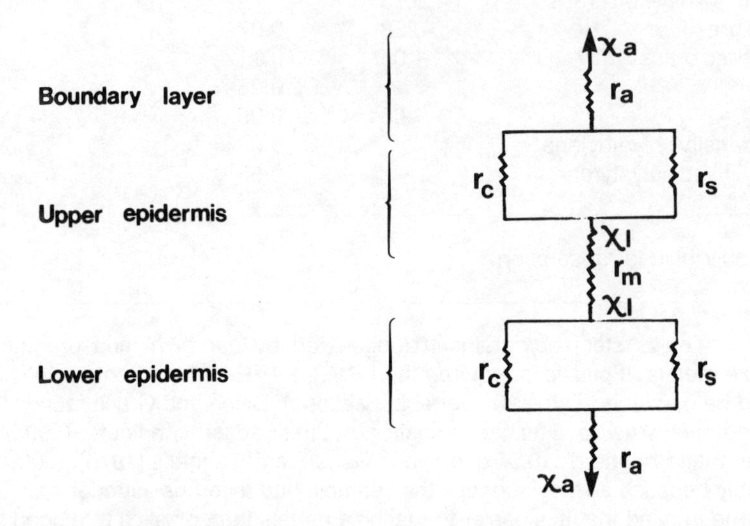

Figure 10.1 Resistance analogue for water vapour diffusion within and outside an amphistomatous leaf.

For amphistomatous leaves the upper and lower halves are considered as two resistance networks in parallel connected by an intercellular resistance (r_m) which is generally small (Jarvis *et al.*, 1967). Whilst this concept is acceptable for water vapour, errors arise when the water vapour resistances are converted to those for carbon dioxide, and the upper and lower surfaces are significantly different (Gale and Poljakoff-Mayber, 1967).

With the flux density of water vapour expressed in g m^{-2} s^{-1} and the concentrations in g m^{-3} (or mol m^{-2} s^{-1} and mol m^{-3}) the units of resistance are s m^{-1}. Originally porometer results were expressed in arbitrary resistance units and only later in absolute units, generally s cm^{-1}. However, because water flux density is proportional to conductance and stomatal conductance (g_s) increases with photon irradiance and humidity so results are best expressed as conductances. In discussion however, resistances are easier to understand (Burrows and Milthorpe, 1976). The units of conductance generally used are mm s^{-1}, or m s^{-1}. Though with a range of 0.5 to 100 mm s^{-1} covering cuticular to boundary layers, the former appears to be the most suitable unit.

It is imperative that where average resistance values are quoted they have been calculated on their reciprocals, otherwise the larger resistances will predominate and the average resistance will be too large. The same also applies to other statistical analyses because the results depend on the measurement of flux densities, so for conductance the variance is reasonably constant whilst for resistance it is proportional to the mean.

To derive resistance we have used

$$E = \Delta\chi / r \qquad 10.3$$

which can be compared with the Fick's law diffusion equation of

$$E = D\,\mathrm{d}\chi / \mathrm{d}x \qquad 10.4$$

where D is the diffusion coefficient and $\mathrm{d}\chi/\mathrm{d}x$ the concentration gradient. The comparison shows that the resistance is inversely proportional to the diffusion coefficient of water vapour.

The diffusion coefficient is both temperature and pressure dependent such that:

$$D = D_o (T/T_o)^{1.75} (P_o/P) \qquad 10.5$$

where D_o is the diffusion coefficient at temperature T_o and pressure P_o, and D the value at T and P. For water vapour at 293 K and 0.1 MPa, D is 24.7 mm^2 s^{-1} (Fuller *et al.*, 1966).

In Britain, with a normal atmospheric pressure range at sea level of 0.0975 to 0.1025 MPa and a range of measurement temperatures from 293 to 308 K, D would range from –8 to +12% of the above value.

Cowan (1977) suggested that if E was expressed in mol m^{-2} s^{-1} and $\Delta\chi$ in mol (water vapour) mol^{-1} (air) (for practical purposes the mole fraction is identical to the volume fraction i.e. vpm), then resistance is in m^2 s mol^{-1}. Because air density is directly proportional to pressure and inversely to temperature, then the effect of pressure on D would be exactly compensated and that of temperature partly so. In the temperature range 278 to 308 K the residual effect of temperature would give rise to variation of –3 to +4%.

Though the scheme has merit, especially as it would unify the units of measurement of biochemists and physiologists, there is still the residual temperature effect and quoting resistance in s mm^{-1} with their associated measurement conditions or perhaps corrected to standard conditions (273 K and 0.1 MPa) is also acceptable.

Resistances expressed in the older units (r) may be converted to the new (r') using the following relation:

$$r' = r V_o P_o T / P T_o \qquad 10.6$$

where V_o is the molar volume of air at T_o and P_o (at 273 K and 0.1 MPa, V_o = 22.7 10^{-3} m^3 mol^{-1}). With r = 0.1 s mm^{-1} then $r' \simeq$ 2.4 m^2 s mol^{-1} (Hall, 1982).

Types of porometers

Measurements independent of water vapour fluxes

Mass flow porometers. This was the first type of porometer (Darwin and Pertz, 1911). It depends on the measurement of the rate of viscous flow of gas through the leaf into or out of a cup attached to the leaf surface, due to an applied pressure gradient. It is most effective for amphistomatous leaves. The main path of air flow is through the stomata on one side, across the intercellular spaces and out of the stomata on the other surface. Thus the volume flow rate depends on the series resistance of the two epidermes and the intercellular spaces. Though the latter resistance is generally small and constant, the technique will only give accurate results if the upper and lower stomata offer similar resistances and respond similarly. Otherwise the results are dominated by the larger resistance.

The major problem with the technique is that the viscous flow resistance through the stomata is measured whereas water vapour exchanges are largely diffusive and include both stomatal and cuticular exchanges. Theoretical attempts have been made to relate the two (Jarvis, 1971) and recently measurements have been made on the same leaves using both mass-flow and dynamic diffusion porometers which show that a relation can only be established for a specific species. Previous work had suggested that g_s was proportional to k^n where k is the viscous flow conductance (mm s^{-1} kPa^{-1}) and n has a value of about 0.4. However, after replotting, the data of Downey *et al.* (1972) indicate a value for n of 0.82 for maize, that of Fischer *et al.* (1977) 0.33 for wheat, and, Milburn's (1979) 0.72 for *Ricinus*.

Many of the recent porometers have been improvements on earlier types (Gregory and Pearse, 1934; Alvim, 1962; Weatherley, 1966;Downey *et al.*, 1972; Fischer *et al.*, 1977; Shimshi, 1977). Others have incorporated electrical transducers to allow continous monitoring (Jodo, 1970; Ogawa and Shibata, 1973; Sheriff and McGruddy, 1976). The only completely new type is that of Milburn (1979). A chamber attached to a 0.1 cm^3 hypodermic syringe is pressed firmly to the leaf surface. The rapid pull back of the syringe plunger creates a vacuum and air is drawn through the leaf into the hypodermic syringe. The plunger is then released, when it returns to a position that depends on how much air has entered. The instrument is calibrated against perforated plates whose viscous conductance can be determined under steady-state conditions.

The problems with the instrument are getting a good seal onto the leaf surface and ensuring reproducible timing of the pull-back and release of the plunger on which the measurement critically depends. With manual timing of the recommended three second period the error is probably large. Another problem is that when the leaf is subject to a large pressure difference it may distort which can influence stomatal opening (Kaufmann and Eckard, 1977). Milburn found that as long as the leaf was well supported this did not occur.

The accuracy of mass flow porometers is irrelevant as the results do not readily relate to diffusive conductances. They are useful for comparisons and, because of their sensitivity to small changes in stomatal aperture, for studying stomatal behaviour. Then it is precision that is important. Few papers quote the necessary figures but considering the variables involved (time, $< 1\%$; pressure, $< 2\%$; volume flow $< 1\%$), the precision should be better than $\pm 5\%$. However, this is critically dependent upon a perfect seal at the leaf surface.

Gas diffusion porometers. These depend on the measurement of the diffusion rate of gases other than water vapour from one side of the leaf to the other. In the past they have used hydrogen (Gregory and Armstrong, 1936), nitrous oxide (Slatyer and Jarvis, 1966), helium (Gale *et al.*, 1967) and radioactive argon (Moreshet *et al.*, 1968). Moreshet and Falkerflug (1978) describe a small portable porometer that measures the diffusion rate of krypton (^{85}Kr) across amphistomatous leaves. Geiger tubes in a supply chamber above and a receiver below the leaf monitor the concentrations of krypton. The time is recorded for the concentration in the lower chamber to change by a fixed proportion of that in the upper. The instrument is calibrated against plates containing holes of calculated conductance (see page **179**). The measured temperature coefficient was 0.37% K^{-1} and the standard error of the results was about $\pm 3\%$ of the mean.

This type of porometer measures the upper epidermal, intercellular and lower epidermal resistances in series. By epidermal stripping the authors show that intercellular resistance does not exceed 7% of the total resistance. In common with the mass flow porometer, differential changes between upper and lower stomata are a problem, as demonstrated in a comparison with a condensation porometer (see page **180**), when at low conductances there is no longer a linear relation. Another possible source of error is that the cuticular permeability may differ between the measurement gas and water vapour. Day (1977) has also shown that it is difficult to equate the diffusion of other gases to that of water vapour, especially at small stomatal opening.

Tritium diffusion porometer. This type of porometer relies on the measurement of the rate of diffusion of tritiated water into the leaf. So far, it has always been combined with photosynthesis measurements using labelled carbon dioxide (Adams *et al.*, 1977; Johnson *et al.*, 1979). Dry air containing $^{14}CO_2$ is bubbled through chilled tritiated water before entering the cuvette. After exposure of the leaf to the air for a known time the leaf is detached and the quantity of ^3H and ^{14}C assayed.

The method assumes that over the time involved, the leaf tissue represents an infinite sink for tritiated water, with a tritium concentration of zero. As long as the sites for evaporation of water within the leaf are not localised, then the tritiated water taken up, and the normal water being lost, will follow identical paths, but in opposite directions. The authors show that

$$(r_l+r_a)^{-1}=F_t/\chi_t \qquad 10.7$$

where F_t is the flux density of tritium into the leaf measured by the radioactive disintegration rate of ^3H in plant tissue per unit exposure time and χ_t is the average concentration of tritium in the air outside the leaf, measured by the radioactive disintegration rate per unit volume.

The boundary layer resistance (r_a) can be determined from the uptake by wet filter paper when r_l is zero. There will be small corrections for differences in diffusion coefficients and a correction for interactions between diffusing gases (Jarman, 1974).

There was good agreement between a gravimetric determination of water loss and that derived from tritium measurements. This is a useful new method of porometry whose accuracy and precision depends entirely upon those of the assay method used.

Measurements dependent on water vapour fluxes

Dynamic diffusion porometers. These consist of a chamber, containing a humidity sensor, sealed onto the leaf surface. In some, dry air is at first admitted to the chamber, to dry it down to a lower starting humidity (van Bavel *et al.*, 1965). In others, the chamber starts from the existing humidity (Kaufmann and Eckard, 1977). In all, the chamber humidity increases, due to leaf transpiration, and the time taken for the humidity to change by a fixed amount (the transit time δt) is recorded. Some have fully automatic operation with a repetitive drying/wetting time cycle, others contain a fan to circulate the air within the chamber (ventilated type).

The theory of this type of porometer is apparently simple.

$$E = V\delta\chi/\delta t \qquad 10.8$$

where E is the mean flux density of water vapour from the leaf into the chamber of volume V during the the transit time δt and $\delta\chi$ is the change in the chamber absolute humidity.

Also

$$E = A(\chi_l - \overline{\chi_a})/(r_l + r_p) \qquad 10.9$$

where A is the area of leaf exposed in the chamber and χ_l and $\overline{\chi_a}$ are the leaf and the average air water vapour concentration ($\overline{\chi_a}$ should be calculated from the average of the logarithms of the lower and upper chamber humidities but where $\delta\chi$ is small and $\overline{\chi_a}$ is large then the linear average is acceptable (Jarvis, 1971)). The resistance r_p is a porometer characteristic which in an ideal instrument would be a diffusion resistance dependent on porometer geometry. In a ventilated instrument r_p should be small. From equations 10.8 and 10.9

$$r_l = (A\delta t(\chi_l - \overline{\chi_a})/V\delta\chi) - r_p \qquad 10.10$$

Because of water vapour absorption by the chamber walls and humidity sensor, the effective chamber volume is always larger than that determined by chamber dimensions (Stigter, 1974; Gandar and Tanner, 1976). This was especially true in early models which used acrylic materials for the chamber and lithium chloride elements for the humidity sensor. From our own observations chambers using stainless steel and Vaisala elements still have an effective volume double their physical one. Results from early models were influenced by the storage humidity, again reflecting the properties of the construction materials (Morrow and Slatyer, 1971a).

Another factor that results in deviation from the theoretical behaviour is the finite response time of the sensors. Under dynamic operating conditions the chamber humidity is not closely followed. Again this was more marked with lithium chloride and polystyrene elements than with Vaisala sensors.

Simple theory predicts that if leaf and sensor temperatures are similar, and with a perfect relative humidity sensor, then the effects of temperature on transit time should be mediated through changes in r_p due to temperature effects on the diffusion coefficient. However, all diffusion porometers show marked temperature effects because the sensors do not have a perfect relative humidity response and the water vapour absorption varies with temperature.

Therefore, because of all these deviations from theory, it is essential to calibrate diffusion porometers against evaporating standards over the anticipated temperature range. The first standards were a series of cylinders of varying length interposed between wet filter paper and the cup (van Bavel *et al.*, 1965). The cylinder resistances were calculated from diffusion theory. Cylinders are not suitable standards, for during the drying cycle, there will be significant penetration of the dry air into the cylinder. They cannot therefore be treated as a steady-state system which diffusion theory requires. The usual calibration method now is to use a plate containing pores, that separates the wet filter paper from the chamber (Kanemasu *et al.*, 1969; Stigter, 1972). Care has to be taken that water from the filter paper does not enter the pores, that the plate temperature is never less than that of the filter paper, when there would be condensation in the pores, and that evaporation from the filter paper elsewhere than into the cup is prevented or the plate temperature will be too low (McCree and van Bavel, 1977). When a graph is plotted of calculated plate resistance against transit time, because of the porometer resistance, r_p, it has an intercept such that there is a finite transit time with zero applied resistance. The magnitude of r_p can be found by extrapolation.

The plate resistance per unit area (R) was originally calculated from the formula given by Brown and Escombe (1900):

$$R=(d/\pi a+1/4)/aDN \qquad 10.11$$

where a and d are the pore radius and depth respectively, N is the number of pores per unit area and D is the diffusion coefficient of water vapour in air. The first term inside the brackets is the basic pore resistance, the second term is the end correction to allow for diffusion away from the pore. Holcomb and Cooke (1977) suggested that the correction was too large at close pore spacing and derived:

$$R=(d/\pi a+1/4-a(N/\pi)^{0.5})/aDN \qquad 10.12$$

Recently Chapman and Parker (1981) derived yet another correction term to the basic pore resistance which requires the use of a table in their publication. The relative magnitudes of the corrections are shown in Table 10.3. Where efforts have been made to test the formula against measurements agreement has been good (Gresham *et al.*, 1975).

In preference to plates with relatively large pores, porous membranes have been used. Their resistance can either be calculated or determined from measurement of the evaporation rate from wet filter paper covered by the membrane under known

Table 10.3 Calculated resistances of plates with pores based on a pore radius and depth of 0.52 and 1.52 mm respectively, with the diffusion coefficient of water vapour in air at 25.2 $mm^2 s^{-1}$.

Pore density cm^{-2}	Pore resistance $s\ mm^{-1}$	End correction $s\ mm^{-1}$ Brown & Escombe	Holcomb & Cooke	Chapman & Parker
4.0	1.775	0.477	0.365	0.387
7.5	0.947	0.254	0.173	0.179
15	0.473	0.127	0.069	0.073
30	0.237	0.063	0.023	0.024

conditions (Stigter, 1974). Another method is to measure the average rate of evaporation from wet filter paper exposed in the chamber. Differing evaporation rates are obtained by varying the exposed area. Water is fed to the paper from a calibrated capillary and the movement of the meniscus is a measure of the evaporation (Korner and Cernusca, 1976; Kaufmann and Eckard, 1977). Both these methods are also applicable to ventilated and steady-state porometers.

Even though calibrations are done at a range of temperatures, temperature differences between leaf and sensor still introduce errors. Early dynamic diffusion porometers were subject to large errors due to convection within the chamber. This led to the advice that both leaf and chamber should be shaded (Morrow and Slatyer, 1971b). The effect is minimised by making the leaf chamber too small for convection (Monteith, 1973) or stirring the air in the cup (ventilated diffusion porometers, Byrne *et al.*, 1970). Another problem arises from the fact that the porometer calibration is done under a given thermal condition and only applies under that condition. With opaque chambers of good thermal conductivity and as long as the leaf has time to equilibrate, then this condition will be reproduced. With transparent chambers, especially under high radiation, the thermal regime cannot be the same as during calibration and the results will be in error. Shading minimises this error.

The condensation porometer described by Moreshet and Yocum (1972) is basically a diffusion porometer with the humidity sensor replaced by a surface kept at 273 K by an ice/water mix. The porometer chamber is flushed with dry air and the time taken from the end of flushing to the onset condensation, on this surface, is the transit time. The major error is in fixing the initial humidity.

Comparisons have been made between dynamic diffusion porometers calibrated by perforated plates and steady-state continuous flow porometers (see page 183) (Johnson, 1981; Bell and Squire, 1981; Norman, pers. comm.) and a tritium diffusion porometer (Johnson *et al.*, 1979; page 177). Where necessary the published conductances have been converted to resistances. Resistance measurements have also been made on layers of Celgard microporous polypropylene film (Amcel Europa SA, Brussels, Belgium) above wet filter paper (Day and Parkinson, unpublished). All these results (Table 10.4) show a linear relation ($y = ax + b$) between resistance measurements made with the different porometers. Where x is the diffusion porometer

Table 10.4 Values for the slopes and intercepts of the equation $y = ax + b$ linking the results from dynamic diffusion porometers (x) and from other porometers (y).

Source	Dynamic diffusion porometer	Comparison porometer	a	b s mm^{-1}
Johnson *et al.* (1979)	Van-Bavel type	Tritium diffusion	0.82±0.16	−0.05 ±0.12
Bell and Squire (1981)	Delta-T	Leeds	0.62±0.05	+0.022±0.01
Johnson (1981)	[1]Wren Instruments	[2]Interface instruments	0.12±0.02	+0.08 ±0.03
Norman (pers. comm.)	[3]Delta-T MKII	[4]LI-COR LI-1600	0.96±0.04	+0.006±0.02
Day and Parkinson (unpubl.)	[5]Crump model 502	Rothamsted	0.80±0.02	−0.05 ±0.02
Day and Parkinson (unpubl.)	LI-COR LI-60	Rothamsted	0.78±0.05	+0.13 ±0.03

[1]Wren Instruments, Hamden, Connecticut, U.S.A.
[2]Interface Instruments, Corvallis, Oregon, U.S.A.
[3]Delta-T, Burwell, Cambridge, U.K.
[4]LI-COR Inc., Lincoln, Nebraska, U.S.A.
[5]T. & J. Crump, Rayleigh, Essex, U.K.

value and y is the other porometer value, then, with the exception of Norman's measurement with the Delta-T MkII porometer, the slope, a, is significantly less than 1.0 i.e. the dynamic diffusion porometer always indicates a greater resistance change. The intercept, b, may be positive or negative. Though the result could be due to errors in calibration of the other types of porometers it is confirmed by other methods (Table 10.5) and also by Dwelle *et al.* (1981) in another comparison with a tritium diffusion porometer. Their results show that with a decrease in measured resistance from 0.5 to 0.15 s mm^{-1}, a more than threefold increase in conductance, the uptake of tritiated water increased by only 50%.

Because the calculated *diffusion* resistance of the plates is accurate, one can only conclude that for many designs, the vapour exchanges within the the porometer chamber are not entirely diffusive; perhaps air movement from the drying cycle has not ceased during the measurement or convection is a problem.

Dynamic-diffusion porometers are small, portable, very quick responding and self-contained, ideal features for field instruments. However, from the above results it is evident that for many of the non-ventilated types calibrated with perforated plates, their accuracy is impossible to determine. With automatic porometers with electronic timing, it will be the precision of the humidity measurement that is limiting. Since the measurements are always made on a wetting cycle hysteresis is irrelevant and the limiting factor is probably again electronic. The results with Celgard suggest a precision of about 10%.

Table 10.5 Resistance for unit 'Celgard' layer determined from the gradients of graphs plotting resistance against layer number.

Method	Resistance per unit layer s mm^{-1} at 293 K
Resistance determined from the evaporation rate measured by weighing	
1. 'Celgard leaves' placed in an unstirred box containing Drierite	0.077±0.008
2. 'Celgard leaves' placed in an open cuvette system with water vapour concentrations measured with a dew-point meter	0.065±0.008
Steady-state constant flow porometer	
Rothamsted type	0.067±0.002
Dynamic-diffusion porometer	
1. Crump model 502	0.087±0.004
2. LI-COR model LI-60 + LI-20S	0.106±0.004

Steady-state diffusion porometer. This consists of a stainless steel chamber with a leaf aperture and at the opposite end, metal gauze (200 mesh), behind which dry air re-circulates (Livingston, 1980). Chamber dimensions are too small for convection. Part way between the leaf surface and the gauze there is a Vaisala humidity sensor (Fig. 10.2). When the chamber is placed on the leaf, a diffusion gradient is set up between the saturated surfaces inside the leaf and the dry air behind the gauze.

Therefore $F=(\chi_l-\chi_p)/(r_l+r_{p1})=\chi_p/r_{p2}$ 10.13

So $r_l=r_{p2}(\chi_l/\chi_p)-(r_{p2}+r_{p1})$ 10.14

where χ_p is the absolute humidity measured by the humidity sensor, r_{p1} is the leaf surface to sensor resistance and r_{p2} the sensor to dry air resistance.

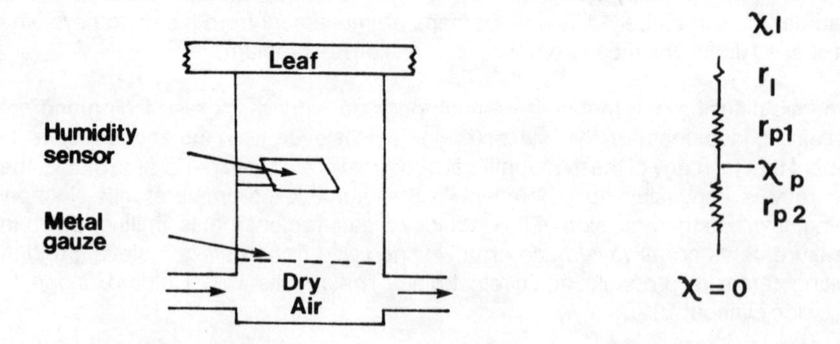

Figure 10.2 The steady-state diffusion porometer and its resistance analogue.

The theory is useful in choosing suitable chamber dimensions and the sensor position. However, the sensor has a finite size whereas theory assumes humidity is meaured at a point, so to determine accurate chamber resistances it is necessary to calibrate against known resistances. Both tubes and perforated plates are acceptable standards under steady-state conditions. This calibration is only necessary initially and subsequently only a humidity calibration is required.

If we assume isothermal conditions then we can write:

$$r_l = r_{p2}(1/h_p) - (r_{p2} + r_{p1}) \qquad 10.15$$

where h_p is the sensor relative humidity. Under normal conditions the leaf and air are not isothermal but are in thermal equilibrium and the necessary correction can easily be derived from a consideration of the energy balance of the leaf (see page **184**).

The accuracy of the porometer will depend on the initial calibrations against standard resistances, and on sensor hysteresis. Precision will largely depend on sensor hysteresis but also on the chamber dimensions and sensor position which determines dh_p/dr_l.

Steady-state continuous flow porometer. This consists of a ventilated chamber placed on the surface or enclosing the leaf into which air of water vapour concentration χ_i (g m^{-3}) is passed at a known volume flow rate Q (m^3 s^{-1}). If the air within the chamber is vigorously stirred then the water vapour concentration of the air leaving the chamber χ_a (g m^{-3}) is representative of that in the chamber air and the transpiration rate is:

$$E = (\chi_a - \chi_i)Q \qquad 10.16$$

but $$E = A(\chi_l - \chi a)/(r_l + r_a) \qquad 10.17$$

From equations 10.16 and 10.17,

$$r_l = A((\chi_l - \chi_a)/(\chi_a - \chi_i))/Q - r_a \qquad 10.18$$

This is the general equation applicable to 'open' cuvette systems for determining leaf resistance. A leaf temperature measurement is required to determine χ_l and if a relative humidity sensor is used then a measurement of sensor temperature is also required. Poor positioning of temperature sensors and small inaccuracies in measurement can lead to large errors in the determination of r_l (Farquhar and Raschke, 1978). For porometry the incoming air is usually dry otherwise χ_i has to be measured.

If we assume isothermal conditions then chamber relative humidity, $h_c = \chi_a/\chi_l$, and with $\chi_i = 0$,

then $$r_l = A(1/h_c - 1)/Q - r_a \qquad 10.19$$

The chamber relative humidity is generally measured by a Vaisala sensor and in the case of the constant flow type (Day, 1977) r_l is calculated from equation 10.19. With the original null-balance instrument (Beardsall *et al.*, 1972) the flow rate is controlled to maintain a constant humidity of 50% when from equation 10.19,

$$r_l = A/Q - r_a \qquad 10.20$$

Other null-balance porometers have controlled the flow to maintain the initial chamber humidity (Bingham and Coyne, 1977).

Conditions within the chamber are rarely isothermal though the system is in a steady-state. The temperature difference can be calculated from the energy balance equation:

$$\rho c_p(T_a - T_l)/r_H + \sigma(T_{ch}^4 - T_l^4) + \alpha I + \lambda(\chi_a - \chi_l)/(r_a + r_l) = 0 \quad 10.21$$

sensible heat	longwave radiation	shortwave radiation	latent heat

where T_l, T_a and T_{ch} are the leaf, air, and chamber wall temperatures respectively, ρ is the air density, c_p its specific heat, σ is Stefan's constant, I is the total short wave irradiance incident on the leaf and α the fraction absorbed, $\lambda$is the latent heat of vaporisation of water and r_H the total resistance to heat transfer to the leaf. In applying this, care must be taken with each component over the area of leaf involved. The above assumes a leaf exposed on one side only.

If we substitute from equation 10.18 for the latent heat term, setting χ_i to zero and to a good approximation $\sigma(T_{ch}^4 - T_l^4) = 4\sigma T_a^3(T_a - T_l)$ (Monteith, 1973) then

$$(T_a - T_l) = (\lambda Q\chi_a/A - \alpha I)/(\rho c_p/r_H + 4\sigma T_a^3) \quad 10.22$$

But the error in r_l resulting from the isothermal assumption is (Parkinson and Day, 1980):

$$\delta r_l = A\triangle(T_a - T_l)/Q\chi_a \quad 10.23$$

with $\triangle = d\chi_{sat}/dT$ where χ_{sat} is the saturation humidity at the average of leaf and air temperature. Substituting for $(T_a - T_l)$ from equation 10.22 then,

$$\delta r_l = (\triangle\lambda - \triangle A\alpha I/Q\chi_a)/(\rho c_p/r_H + 4\sigma T_a^3) \quad 10.24$$

But in a well stirred chamber $4\sigma T_a^3$is an order of magnitude smaller than $\rho c_p/r_H$ and in an opaque chamber $I = 0$ then,

$$\delta r_l = \triangle\lambda r_H/\rho c_p \quad 10.25$$

With small error $\triangle$ can be determined at T_a when the correction is independent of leaf resistance r_l but increases with temperature. It is still necessary to measure air temperature but an accuracy of one Kelvin will give only 5% error on the correction (this is the temperature coefficient of $\triangle$).

The use of a relative humidity measurement and the calculation of r_l from equations 10.19 and 10.24, because it averages over the whole leaf area, is likely to be more accurate than measuring both leaf and sensor temperatures and then applying equation 10.18.

The boundary layer resistance, r_a, is determined by measuring the resistance of wet filter paper and r_H from the temperature gradient between the paper and sensor. However, r_H can be determined more accurately from measurements on a series of Celgard standards.

The advantage of steady-state over the other types of porometer is that once r_a and r_H have been established then the only calibrations necessary are those of volume flow rate and relative humidity. Initially this type of instrument was cumbersome, requiring gas cylinders, but the latest types are small with self-contained air pumps. Null-balance instruments have the advantage over the constant flow in always working at one humidity when water absorption is no longer a problem. Also the humidity operation can be ambient so that the leaf is not subjected to low humidities at small stomatal opening which may induce further stomatal closure.

Recently steady-state porometers have been combined with CO_2 flux measurements to provide complete photosynthesis/transpiration measuring sytems (Bingham and Coyne, 1977; Griffiths and Jarvis, 1981; Bell and Incoll 1981a; Schulze *et al.*, 1982).

Some porometers use mass-flow meters to measure the rate of chamber air flow and then it is advantageous to express the resistance in m^2 s mol^{-1} as it eliminates the measurement of atmospheric pressure and it is only necessary to correct for the residual effects of temperature (see page 173), then

$$r_{l(273K)} = r_{measured}(T_a/273)^{0.75} \qquad 10.26$$

Other porometers use rotameters or capillary flowmeters where, for a constant pressure differential, the mass flowing is inversely proportional to $\rho^{0.5}$ and the correction is then,

$$r_{l(273K,0.1MPa)} = r_{measured}(T_a/273)^{1.25}(0.1/P_a) \qquad 10.27$$

Where steady-state continuous flow porometers have been compared with rates of water loss determined gravimetrically, the agreement has been good (Day, 1977; and Table 10.4). The accuracy depends to some extent on the magnitude of r_l (Bell and Incoll, 1981b) but it is about ±10%. The precision should be considerably better than that as it depends only on the precision of the volume flow (±1%) and humidity measurement (±2%).

Measurement of evaporation from animal epidermis

In one method, chambers containing drying agents were placed on the skin surface and the increase in weight of the desiccant measured. The alternative method was similar to steady-state porometry (see page 183). Both modify the natural humidity gradient above the skin and therefore the evaporation rate; the evaporimeter (Nilsson, 1977) was an attempt to overcome this. Nilsson initially showed that above an evaporating surface in still air there is a region about10 mm deep in which the humidity gradient is constant and related to the evaporation rate. His evaoporimeter consists of a cylinder (length 15.5 mm; diameter 12.5 mm) with one end placed on the skin surface and the other open to the atmosphere. These dimensions were chosen so as to maintain the natural boundary layer and eliminate convection. Along the length of the cylinder, two humidity and their associated temperature sensors are placed 4 mm apart, to measure the humidity gradient. The instrument is calibrated against gravimetrically-determined evaporation rates and has been extensively used in the study of water loss from the skin of newborn infants (Rutter, 1980).

Wheldon and Monteith (1980) showed that under certain conditions the instrument under-estimated evaporation and derived the appropriate corrections. Under still air conditions, where it is normally used, its accuracy, estimated from Wheldon and Monteith (1980), is about ±3% with a precision of about ±10% of the evaporation rate. However, any wind would disturb the uniformity of the gradient and affect the readings.

Conclusions

Porometers which do not measure water vapour are generally best avoided because of the difficulties in equating their results to values for water vapour. Where suitable measuring equipment is already available then the tritium diffusion porometer could be used although the results are not given immediately.

The non-ventilated dynamic diffusion porometer is a useful instrument though there is a question mark over some models as to their accuracy when calibrated with perforated plates. The ventilated type is calibrated by other methods and should be accurate.

The steady-state diffusion porometer is an exciting new development and merits further investigation.

Steady-state continuous flow porometers have the advantage of requiring only air flow rate and humidity calibrations. The null-balance type although more complicated operates at a selected humidity which may be more suitable if leaves respond very rapidly to humidity (<15 s).

If a porometer is required to completely enclose a leaf or a branch with leaves, only ventilated porometers are suitable because the position of the evaporating surfaces within the chamber cannot be closely defined.

Acknowledgements

The author would like to thank Dr N. Rutter for information on the evaporimeter, Professor K. M. King for details of the steady-state diffusion porometer and Professor J. Norman for results from comparisons of porometers.

References

Adams, J.A., Johnson, H.B., Bingham, F.T. & Yermands, D.M. (1977). Gaseous exchange of *Simmondsia chinesis* (Jojoba) measured with a double isotope porometer and related to water stress, salt stress and nitrogen deficiency. Crop Sci., **17**, 11-15.

Alvim, P. de T. (1962). A new type of porometer for measuring stomatal opening and its use in irrigation studies. Symp. Methodology Eco-Physiology. UNESCO, Montpellier, France. pp.325-329.

van Bavel, C.H.M., Nakayama, F.S. & Ehrley, W.L. (1965). Measuring transpiration resistance of leaves. Pl. Physiol., **40**, 535-540.

Bell, C.J. & Incoll, L.D. (1981a). A handpiece for the simultaneous measurement of photosynthetic rate and leaf diffusive conductance. I. Design. J. exp. Bot., **32**, 1125-1134.

Bell, C.J. & Incoll, L.D. (1981b). A handpiece for the simultaneous measurement of photosynthetic rate and leaf diffusive conductance. II. Calibration. J. exp. Bot., **32**, 1135-1142.

Bell, C.J. & Squire, G.R. (1981). Comparative measurements with two water vapour diffusion porometers (dynamic and steady-state). J. exp. Bot., **32**, **1143-1156.**

Beardsell, M.F., Jarvis, P.G. & Davidson, B. (1972). A null-balance diffusion porometer suitable for use with leaves of many shapes. J. appl. Ecol., **9**, 677-690.

Bingham, G.E. & Coyne, P.I. (1977). A portable, temperature-controlled, steady-state porometer for field measurements of transpiration and photosynthesis. Photosynthetica, **11**, 148-160.

Bloom, A.J., Mooney, H.A., Bjorkman, O. & Berry, J. (1980). Materials and methods for carbon dioxide and water exchange analysis. Plant, Cell & Environ., **3**, 371-376.

Brown, H.T. & Escombe, F. (1900). Static diffusion of gases and liquids in relation to the assimilation of carbon and translocation in plants. Phil. Trans. R. Soc. Lond. Ser. B., **193**, 223-291.

Burrows, F.J. & Milthorpe, F.L. (1976). Stomatal conductance in the control of gas exchange. *In* Water Deficits and Plant Growth. Vol. IV. Soil water measurement, plant responses and breeding for drought resistance, ed. T.T. Kozlowski, pp.103-152. New York, London: Academic Press.

Byrne, G.F., Rose, C.W. & Slatyer, R.O. (1970). An aspirated diffusion porometer. Agric. Meteorol., **7**, 39-44.

Chapman, D.C. & Parker, R.L. (1981). A theoretical analysis of the diffusion porometer: Steady diffusion through two finite cylinders of different radii. Agric. Meteorol., **23**, 9-20.

Cowan, I.R. & Milthorpe, F.L. (1968). Plant Factors Influencing the Water Status of Plant Tissues. *In* Water Deficits and Plant Growth. Vol. I. Development, control and measurement, ed. T.T. Kozlowski, pp.137-193. London & New York: Academic Press.

Cowan, I.R. (1977). Stomatal behaviour and environment. *In* Advances in Botanical Research, Vol. 4, eds. R.D. Preston and H.W. Woolhouse pp.117-228. London & New York: Academic Press.

Crank, J. & Park, G.S. (1968). Diffusion in Polymers. London & New York: Academic Press.

Darwin, F. & Pertz, D.F.M. (1911). On a new method of estimating the aperture of stomata. Proc. R. Soc. Lond. Ser. B, **1**, 136-154.

Day, W. (1977). A direct reading continuous flow porometer. Agric. Meteorol., **18**, 81-89.

Dixon, M. & Grace, J. (1982). Water uptake by some chamber materials. Plant, Cell & Environ., **5**, 323-327.

Downey, L.A., Anlezark, R.N. & Muirhead, W.A. (1972). Construction, calibration and field use of rapid-reading viscous flow porometer. J. appl. Ecol., **9**, 431-437.

Dwelle, R.B., Kleinkope, G.E., Steinhorst, R.K., Pavek, J.J. & Hurley, P.J. (1981). The influence of physiological processes on tuber yield of potato clones (*Solanum tuberosum* L.): Stomatal diffusive resistance, stomatal conductance, gross photosynthetic rate, leaf canopy, tissue nutrient levels and tuber enzyme activities. Potato Res., **24**, 33-47.

Farquhar, G.D. & Raschke, K. (1978). On the resistance to transpiration of the sites of evaporation within the leaf. Pl. Physiol., **61**, 1000-1005.

Fischer, R.A., Sanchez, M. & Syme, J.R. (1977). Pressure chamber and air flow porometer for rapid indication of water status and stomatal condition in wheat. Exp. Agric., **13**, 341-351.

Fuller, E.N., Schettler, P.D. & Giddings, J.C. (1966). A new method for prediction of binary gas-phase diffusion coefficients. Ind. Engng. Chem., **58**, 19-27.

Gale, J. & Poljakoff-Mayber, A. (1967). Resistance to gas flow through the leaf and its significance to measurements made with viscous flow and diffusion porometer. Israel J. Bot., **16**, 205-211.

Gale, J., Poljakoff-Mayber, A. & Kahane, I. (1967). The gas diffusion porometer technique and its application to the measurement of leaf mesophyll resistance. Israel J. Bot., **16**, 187-204.

Gandar, P.W. & Tanner, C.B. (1976). Water vapour sorption by the walls and sensors of stomatal diffusion porometers. Agron. J., **68**, 245-249.

Gregory, F.G. & Armstrong, J.I. (1936). The diffusion porometer. Proc. R. Soc. Lond. Ser. B., **121**, 27-42.

Gregory, F.G. & Pearse, H.L. (1934). The resistance porometer and its application to the study of stomatal movement. Proc. R. Soc. Lond. Ser. B., **144**, 477-493.

Gresham, C.A., Sinclair, T.R. & Wuenscher, J.E. (1975). A ventilated diffusion porometer for measurement of the stomatal resistance of pine fascicles. Photosynthetica, **9**, 72-77.

Griffiths, J.H. & Jarvis, P.G. (1981). A null balance carbon dioxide and water vapour porometer. J. exp. Bot., **32**, 1157-1168.

Hall, A.E. (1982). Mathematical model of plant water loss and plant water relations. *In* Encyclopaedia of Plant Physiology (New Series) Vol. 12B, eds. A. Pirson & M.H. Zimmermann, pp.231-261, Berlin, Heidelberg & New York: Springer-Verlag.

Holcomb, D.P. & Cooke, J.R. (1977). Diffusion resistance of porometer calibration plates determined with an electrolytic tank analog. ASAE Mtg., No. 77-5509, St. Joseph, MI.

Jarman, P.D. (1974). The diffusion of carbon dioxide and water vapour through stomata. J. exp. Bot., **25**, 927-936.

Jarvis, P.G. (1971). The estimation of resistance to carbon dioxide transfer. *In* Plant Photosynthetic Production: Manual of Methods, eds. Z. Sestak, J. Catsky & P.G. Jarvis, pp. 599-631. The Hague: Dr W. Junk.

Jarvis, P.G., Rose, C.W. & Begg, J.E. (1967). An experimental and theoretical comparison of viscous and diffusive resistance to gas flow through amphistomatous leaves. Agric. Meteorol., **4**, 103-117.

Jodo, S. (1970). Stomatal movement and water relations in crops. I. Performance test on a newly improved recording porometer. Proc. Crop. Sci. Soc. Japan, **39**, 431-439.

Johnson, H.B., Rowlands, P.G. & Ting, I.P. (1979). Tritium and carbon-14 double isotope porometer for simultaneous measurements of transpiration and photosynthesis. Photosynthetica, **13**, 409-418.

Johnson, J.D. (1981). Two types of ventilated porometers compared on broadleaf and coniferous species. Pl. Physiol. **68**, 506-508.

Kammermeyer, K. (1957). Silicone rubber as a selective barrier. Ind. Engng. Chem., **49**, 1685-1686.

Kanemasu, E.T. (ed.) (1975). Measurement of stomatal aperture and diffusive resistance. ed. E.T. Kanemasu, Bulletin 809, Washington State University: College of Agriculture Research Centre.

Kanemasu, E.T., Thurtell, G.W. & Tanner, C.B. (1969). Design, calibration and field use of a stomatal diffusion porometer. Pl. Physiol., **44**, 881-885.

Kaufmann, M.R. & Eckard, A.N. (1977). A portable instrument for rapidly measuring conductance and transpiration of conifers and other species. For. Sci., **23**, 227-237.

Korner, C. & Cernusca, A. (1976). A semi-automatic, recording diffusion porometer and its performance under alpine conditions. Photosynthetica, **10**, 172-181.

Lebovits, A. (1966). Permeability of polymers to gases, vapours and liquids. Mod. Plast., **43**, 139-213.

Livingston, N. (1980). The development and field use of a steady state diffusion porometer. MSc. degree Thesis. University of Guelph, Ontario Agricultural College.

Major, C.J. & Kammermeyer, K. (1962). Gas permeability of plastics. Mod. Plast., **39**, 135-150.

McCree, K.J. & van Bavel, C.H.M. (1977). Calibration of leaf resistance porometers. Agron. J., **69**, 724-726.

Milburn, J.A. (1979). An ideal viscous flow porometer. J. exp. Bot., **30**, 1021-1034,

Monteith, J.L. (1973). Principles of environmental physics. London: Edward Arnold.

Moreshet, S. & Falkenflug, V. (1978). A krypton diffusion porometer for the direct field measurement of stomatal resistance. J. exp. Bot., **29**, 267-275.

Moreshet, S. & Yocum, C.S. (1972). A condensation type porometer for field use. Pl. Physiol., **49**, 944-949.

Moreshet, S., Stanhill, G. & Koller, D. (1968). A radioactive tracer technique for the direct measurement of the diffusion resistance of stomata. J. exp. Bot., **19**, 460-467.

Morrow, P.A. & Slatyer, R.O. (1971a). Leaf resistance measurements with diffusion porometers: precautions in calibration and use. Agric. Meteorol., **8**, 223-233.

Morrow, P.A. & Slatyer, R.O. (1971b). Leaf temperature effects on measurements of diffusive resistance to water vapour transfer. Pl. Physiol., **47**, 559-561.

Nilsson, G.E. (1977). Measurement of water exchange through skin. Med. Biol. Eng. Comput., **15**, 209-218.

Ogawa, T. & Shibata, K. (1973). A simple porometer for precise recording of leaf resistance. Pl. Cell Physiol., **14**, 1039-1043.

Parkinson, K.J. & Day, W. (1980). Temperature corrections to measurements made with continuous flow porometer. J. appl. Ecol., **17**, 457-460.

Roff, W.J. & Scott, J.R. (1971). Fibres, films, plastics and rubbers. London: Butterworths & Co. Ltd.

Rutter, N. (1980). Water loss from the skin of newborn infants. M.D. Thesis, University of Cambridge.

Schulze, E.D., Hall, A.E., Lange, O.L. & Walz, H. (1982). A portable steady-state porometer for measuring the carbon dioxide and water vapour exchange of leaves under natural conditions. Oecologia (Berl.), **53**, 141-145.

Shepherd, W. (1973). Moisture absorption by some instrument materials. Rev. scient. Instrum., **44**, 234.

Sheriff, D.W. & McGruddy, E. (1976). Changes in leaf viscous flow resistance following excision, measured with a new porometer. J. exp. Bot., **27**, 1371-1375.

Shimshi, D. (1977). A fast-reading viscous flow leaf porometer. New Phytol., **78**, 593-598.

Slatyer, R.O. (1971). Effect of errors on measuring leaf temperature and ambient gas concentration on calculated resistances to CO_2 and water vapour exchanges in plant leaves. Pl. Physiol., **47**, 269-274.

Slatyer, R.O. & Jarvis, P.G. (1966). Gaseous-diffusion porometer for continuous measurement of diffusive resistance of leaves. Science, N.Y., **151**, 574-576.

Slavik, B. (1971). Determination of stomatal aperture. *In* Plant Photosynthetic Production: Manual of Methods. eds. Z. Sestak, J. Catsky & P.G. Jarvis, pp.556-565. The Hague: Dr W. Junk.

Slavik, B. (1974). Methods of studying plant water relations. Prague: Academia.

Stigter, C.J. (1972). Leaf diffusion resistance to water vapour and its direct measurement. 1. Introduction and review concerning relevant factors and methods. Meded. LandbHoogesch. Wageningen, Report 72-73.

Stigter, C.J. (1974). The epidermal resistance to diffusion of water vapour: An improved measuring method and field results in Indian corn (*Zea mays*). Agric. Res. Rep. (Versl. Landbouwk. Onderz.), Report 831.

Thom, A.S. (1968). The exchange of momentum, mass and heat between an artificial leaf and the air flow in a wind tunnel. Q. Jl. R. met. Soc., **94**, 45-56.

Waggoner, P.E. (1966). Moisture loss through the boundary layer. *In* Biometeorology, Vol. 3. Proc. 4th Int. Biometeorol. Cong., New Brunswick, eds. S.W. Tromp & W.H. Wieke, pp.41-52, Amsterdam: Swets & Zeitlinger Publ. Co.

Weatherley, P.E. (1966). A porometer for use in the field. New Phytol., **65**, 376-387.

Wheldon, A.E. & Monteith, J.L. (1980). Performance of a skin evaporimeter. Med. Biol. Eng. Comput., **18**, 201-205.

Woolley, J.T. (1967). Relative permeabilities of plastic films to water and carbon dioxide. Pl. Physiol., **42**, 641-643.

Yasuda, H. & Stone, W. (1966). Permeability of polymer membranes to dissolved oxygen. J. Polym. Sci., **4**, 1314-1316.

Chapter 11: Instruments for measuring plant water potential and its components

G. S. Campbell
Department of Agronomy and Soils, Washington State University, Pullman, WA 99164, U.S.A.

The status of water in plants can be assessed in several ways. Plant appearance, growth rate, and rates of several other physiological processes can be used as indicators of water stress. Relative water content of plant tissue can also be used as an indicator. The discussion will be concerned with the measurement and interpretation of plant water potential and its components.

Water potential and its components

Before discussing methods for measuring water potential, we need to define terms and select appropriate units. Water potential has been defined in both mechanistic and thermodynamic terms. A mechanistic description was given by a soil physics terminology committee of the International Soil Science Society (Aslyng, 1963). They defined the total potential as "the amount of work that must be done per unit quantity of pure water in order to transport reversibly and isothermally an infinitesimal quantity of water from a pool of pure water at a specified elevation at atmospheric pressure to the soil water (at the point under consideration)". In thermodynamic terms, the water potential is the chemical potential, or partial specific Gibbs free energy of the water in the soil or plant system (Slatyer, 1967). Since the energy of water in the soil-plant-atmosphere system is usually lower than that of pure free water, the water potential is usually a negative number.

The unit of energy or work in the International System (SI) is the joule (J). Water potential is the energy or work per unit quantity of water, with the quantity being expressed as mass, volume, weight, or moles of water. The mass basis potential (units of J/kg) is preferred because the amount of water does not change with changes in density due to temperature and pressure, and because the relation to energy status of the water is clear in these units (Campbell and van Schilfgaarde, 1981). Energy per unit volume is equivalent to pressure. In SI, the pressure unit is the pascal (Pa).

Water potentials in the plant and its environment

The total water potential is often expressed as the sum of a number of component potentials. A typical set is

$$\psi = \psi_o + \psi_p + \psi_g + \psi_\Omega \qquad 11.1$$

where ψ represents the potential, and the subscripts o, p, g, and Ω represent osmotic, pressure, gravitational, and overburden components. In systems with negative pressure ψ_p is often replaced by ψ_m, the matric potential (Passioura, 1980). Many of the components are defined by the methods that are used to measure them, and therefore are not completely independent or additive, as is implied by equation 11.1. A proposal to substitute a more fundamental set (Spanner, 1973) requires more difficult experimental methods, and apparently has not been used.

Solutes in living cells, cell walls, and xylem reduce the osmotic potential of plant water. Osmotic potentials in cell walls and xylem typically range from -100 J/kg to near zero, and are often neglected in plant water studies (Boyer, 1967a). Osmotic potentials in cells may range from -300 J/kg in growth chamber grown or understorey plants to -7000 J/kg or -8000 J/kg for some xerophytes. Typical values are around -1000 to -1500 J/kg.

In plant cells, the pressure potential is generally positive, ranging from zero to values large enough to balance the osmotic potential. Within a plant cell

$$\psi=\psi_o+\psi_p \tag{11.2}$$

so, if

$$\psi = 0, \psi_p = -\psi_o.$$

Pressure in xylem is generally negative. Equation 11.2 applies to xylem, as well as other cells, but ψ_o is usually near zero, requiring that the pressure be about equal to the total potential, which can go as low as –7000 or –8000 J/kg, and typically reaches midday values of –1000 to –2000 J/kg. Briggs (1950) has shown that water in capillaries is capable of supporting tensions well beyond these values.

We will use the term "matric" to describe the negative pressure in cell walls and unsaturated soil (Passioura, 1980). The matric potential of cell walls is the same as that for osmotic potential in cells.

The gravitational and overburden components are calculated from

$$\psi_g=gz \tag{11.3}$$

and

$$\psi_\Omega=\kappa P_L/\rho_w \tag{11.4}$$

where g is the gravitational constant, z is the height from a reference pool, κ is the fraction of the applied pressure that is transmitted to the matrix water, P_L is the matrix load pressure, and ρ_w is the density of water. The gravitational potential increases 10 J/kg for each metre of plant height, so is likely to be negligible except for very tall plants. The magnitude of ψ_Ω will be discussed later.

The water content-water potential relation

The work required to remove water from a soil or plant is obviously related to the amount of water in the system. It is often useful to be able to describe the relation – both for inferring one value from the other and as a basis for understanding measurements and sources of error. The osmotic potential-water content relation for the symplastic fluid, assuming it is an ideal solution in which the amount of solute present remains constant is

$$\psi_o = \psi_{ot} C_t / C \qquad 11.5$$

where C is the symplastic water fraction of the tissue, and the subscript, t, refers to the fully turgid condition. No theoretical expression exists for the matric potential-water content relation of the apoplast, but the empirical relation

$$\psi_m = \psi_{mt} (B_t / B)^b \qquad 11.6$$

has been shown to give reasonable fits to the data (Campbell *et al.*, 1979; Acock and Grange, 1981). Here, b is a constant with typical values ranging from 1 to 7 and B is the apoplastic water fraction.

The turgor pressure-water content relation for an intact cell can be approximated by (Warren Wilson, 1967)

$$\psi_p = \epsilon (R - R_o) \qquad 11.7$$

where ϵ is the cell wall elastic modulus, and R is the relative water content of the cell ($R = B + C$). Subscript o indicates the zero turgor value. The elastic modulus, ϵ, is assumed to be a constant, positive number at $R > R_o$ and zero for $R < R_o$. Consequences of less restrictive assumptions about ϵ were explored by Campbell *et al.* (1979) and found to have little effect on the analysis presented here.

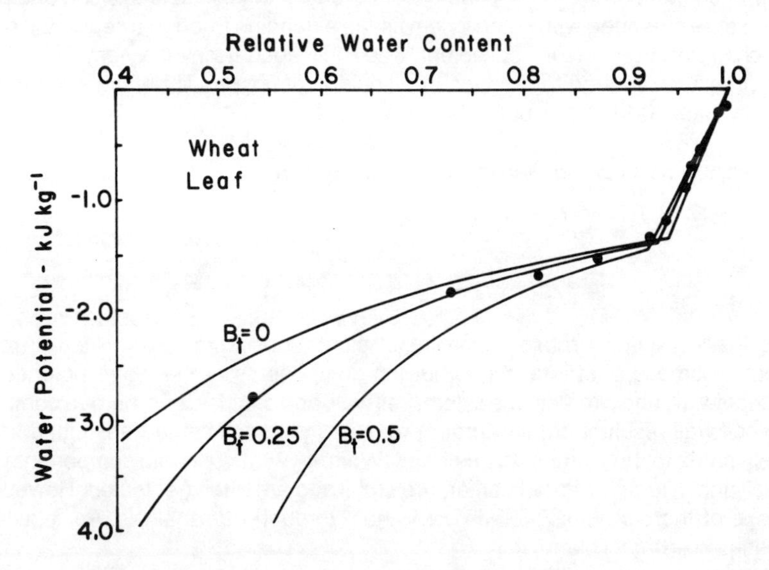

Figure 11.1 Water potential–water content relationship for a wheat leaf. Lines are drawn with equation 11.8 and ϵ = 21.6 kJ kg^{-1}, ψ_{ot} = −1.28 kJ kg^{-1}, B_t = 0, 0.25, or 0.5, $\alpha = B_t/2$, and $R_o = 1 + \psi_{ot}/(\epsilon(1-\alpha))$. Points are measured values for a wheat leaf. Turgor loss occurs around $R = 0.93$.

Since plant tissue is a combination of cells, solute, and matrix, an equation that describes $\psi(R)$ for tissue must combine equations 11.5-11.7. Following Campbell *et al.* (1979) we can write:

$$\psi(R)=\epsilon(1-\alpha)(R-R_o)+\psi_{ot}(1-B_t)/((1-\alpha)R-B_t+\alpha) \qquad 11.8$$

B_t is the apoplastic water fraction at full turgor, ψ_{ot} is the osmotic potential of fully turgid tissue, and α is the fraction of the water removed from the tissue that comes from the apoplast. Fig. 11.1 shows values for $\psi(R)$ for some representative values of the parameters. Important things to note are a) water potentials in the range from full to zero turgor occur over a very narrow relative water content range (1 to 0.93) and b) as the apoplastic water fraction increases, the tissue water capacitance ($dR/d\psi$) of turgid and flaccid tissue becomes more similar with turgid capacitance increasing and flaccid decreasing.

Measurement of tissue water potential with a pressure chamber

Pressure methods require the application of pneumatic pressure to the tissue to increase its water potential. The pressure required to bring the tissue water potential to zero is taken as the negative of the tissue water potential before pressurisation. If we assume that the water potential gradients within a particular organ are small, then, to first approximation, $\psi_{xylem} = \psi_{cell\ wall} = \psi_{cell}$ within the tissue. For a pressure equilibration measurement of tissue water potential, the tissue is excised and placed in a pressure chamber, with some xylem tissue extending through a seal to the outside of the chamber. Pressure is applied until free liquid appears in the xylem. The negative of the pressure required to produce free liquid at atmospheric pressure in the xylem is usually taken as the water potential of the tissue.

An analysis of water potential in the intact plant shows

$$\begin{aligned} \psi^i_{xylem} &= \psi^i_{px} + \psi^i_{ox} \\ \psi^i_{cellwall} &= \psi^i_{\Omega w} + \psi^i_{ow} + \psi^i_{mw} \\ \psi^i_{cell} &= \psi^i_{pc} + \psi^i_{oc} \end{aligned} \qquad 11.9$$

where the subscript, i refers to the intact plant, subscripts p, o, m and Ω refer to pressure, osmotic, matric and overburden components of the water potential, and subscripts x, w, and c refer to the xylem, cell wall, and cell. The overburden component in the cell wall results from the turgor pressure in the cell being transmitted through the cell matrix to the water in the cell wall (Walter, 1963). Osmotic components in the cell wall and xylem, as stated earlier, are small and are often neglected. However, for the sake of thoroughness we will carry them through our analysis. For equilibrium conditions in an intact plant

$$\psi^i_{ox} + \psi^i_{px} = \psi^i_{pc} + \psi^i_{oc} \qquad 11.10$$

When the tissue is excised, the tension in the xylem fluid is released, and water moves from the xylem into the surrounding cells, increasing their turgor. When pressure is applied to the tissue in the pressure chamber, the cell water potential becomes

$$\psi_{cell} = \psi_{pc} + \psi_{oc} + P \qquad 11.11$$

where P is the pneumatic potential, equal to the applied pressure divided by the density of water. The cell wall potential is increased similarily by the applied pressure, but the water filled walls of the xylem vessels form a membrane in the xylem that extends through the pressure chamber seal, and when the meniscus of that fluid is flat, the pressure potential of the xylem, ψ_{px}, must be zero. If the wall and cell potentials are at equilibrium with the xylem, then, at the balance pressure,

$$\psi_{ox} = \psi_{pc} + \psi_{oc} + P \qquad 11.12$$

and

$$\psi_{ox} - P = \psi_{pc} + \psi_{oc} \qquad 11.13$$

The water potential of the tissue when it was part of the intact plant would therefore equal $\psi_{ox} - P$ if the water content of the cells and cell walls after equilibration were the same as that for the tissue when it was part of the plant. Prevention of water loss from the tissue, and restoration of the xylem fluid to its pre-excision position help assure that this condition is met. Equation 11.13 therefore provides a rational basis for tissue water potential measurement using pressure equilibration. The principles involved are similar to those which apply to soil water potential measurement with the pressure plate apparatus (Taylor and Ashcroft, 1972; Passioura, 1980).

An extensive review of pressure chamber methods and results is given by Ritchie and Hinckley (1975). The equipment necessary for making this measurement includes a pressure chamber with a seal appropriate for allowing xylem to extend from the tissue inside the chamber to an observable location outside, and a source of compressed gas (usually air or nitrogen). The pressure chamber operates at high pressure, posing a potential hazard to the operator. Serious injuries have resulted from failure to take adequate precautions. I recommend that only commercial pressure chambers are used. These have been pressure tested, and have several safety features which minimise risks to the operator. Suppliers of commercial units are listed at the end of this chapter.

For single measurements, the gas used in the pressure chamber can be either air or nitrogen. For repeated measurements on the same tissue, nitrogen, or a mixture of oxygen and nitrogen should be used to avoid membrane oxidation. Tyree *et al.* (1978) recommend a mixture of 90-95% pure nitrogen and 5-10% compressed air. Compressed gases are generally supplied in cylinders at pressures around 15 MPa. These also can be very dangerous, if mishandled.

In addition to these two items, it is often helpful to have a strong light source and a low power microscope available to observe the xylem during pressurisation. The pressure chamber can be used for measuring water potential of any plant part, provided xylem tissue can be found which can lead from that part through the pressure chamber seal. The pressure chamber has been used successfully on leaves, tubers, fruits, branches and leaf parts. If the entire organ is too large to fit inside the pressure chamber, it is often possible to cut away part of the sample and make a valid measurement on the remainder. For example, leaves of sunflower (*Helianthus annuus*) or sugar beet (*Beta saccharifera*) are large and have petioles which are difficult to seal in the pressure chamber. Water potential of these leaves can be measured by dissecting out a portion of the leaf with the midrib or a vein attached to extend through the pressure chamber seal.

The requirements for accurate pressure chamber methods are implicit in the assumptions made for the measurement. These are: a) the water content of the cells and cell walls is the same at balance pressure as existed in the intact plant, and b) the total water potential is constant throughout the tissue, and equal to the xylem fluid osmotic potential at the balance pressure. In addition, it is often assumed that the osmotic potential of the xylem fluid is negligible . Boyer (1967a) measured xylem fluid potentials and found them to average –50 J/kg, which is often very small compared to the total potential.

Precautions which help to assure accurate tissue water potential measurements with the pressure chamber are given by Baughn and Tanner (1976b). These include covering the tissue to prevent water loss during the measurement process, slow pressurisation to allow near-equilibrium conditions to be maintained within the leaf, careful observation of the xylem so that pressurisation is halted just when a free fluid surface appears at the cut end of the xylem, and minimising the amount of tissue which extends through the seal so as to minimise the possibility of tissue filling with water under pressurisation which was not filled when the tissue was attached to the plant.

For prevention of water loss, Gandar and Tanner (1975) and Baughn and Tanner (1967b) suggest wrapping the tissue tightly in moist cheesecloth and then in a polyethylene bag prior to excision, and leaving the cloth and bag in place during the measurement. Leach *et al.* (1982) suggested wrapping the leaf with "clingfilm", which they claim to be more effective than the Baughn and Tanner method. Other researchers have tried to reduce tissue water loss by putting free water in the chamber or wrapping tissue with aluminium foil or polyethylene. These methods are not as effective as that of Baughn and Tanner because of the large and rapid temperature changes which occur during pressurisation (Puritch and Turner, 1973) and the fact that dry air is introduced when the chamber is pressurised. Turner and Long (1980) have shown that water loss from leaves during the first 30 s after excision can result in measurement errors as large as 700 J/kg, and that humidifying the chamber had no measurable effect on these errors. An indication of the error one might expect from tissue water loss can be obtained by referring to Fig. 11.1. Note that, in the turgid range, a 1% change in relative water content can result in a 200 J/kg error in the water potential measurement. At lower water potentials, approaching zero turgor, the tissue water capacity is much higher, and errors due to water loss are less serious. This is also shown by Turner and Long (1980). Midday water potential measurements, near zero turgor, may therefore suffer less from failure to prevent water loss from the sample during the measurement, while measurements at higher turgor pressure are more susceptible to error. Errors can arise from using the moist cheesecloth if the cloth touches a cut or damaged portion of the leaf. Then, water is immediately injected into the leaf and causes a false endpoint.

Baughn and Tanner (1976b) used a rate of pressure increase of 20 to 50 J kg^{-1} s^{-1} until the pressure was within 100-200 J/kg of the balance pressure. Then they reduced the rate to 5 J kg^{-1} s^{-1}. More rapid rates of increase may lead to water potential gradients within the tissue. Pressurisation should cease as soon as water appears at the cut xylem end. If overpressurisation occurs, water will be lost from the xylem, and errors will result. As with evaporative water loss, these errors are most serious at high water potentials. False endpoints are possible with some species which exude resins or latex from the tissue surrounding the xylem. Air can bubble through these materials making the point of first water appearance difficult to determine. If such exudates are a problem it is usually possible to peel or dissect away offending tissue, so that the xylem by itself can be obsereved.

After the balance pressure has been attained, one should release the pressure sufficiently to allow the water to recede into the xylem, and then repressurise the chamber. The second reading should agree with the first to within 30 to 50 J/kg. If it does not, a third reading should be taken. If the readings increase on each successive try, water is being lost from the tissue. Generally the pressure will be a bit lower the second time, indicating that the tissue was not at equilibrium during the first reading.

The precision of the pressure chamber, with carefuł technique, is 20 to 50 J/kg, depending on species, water potential, and other variables. The uncertainty comes from difficulty in identifying the end point, water loss, and changes in tissue water distribution. Accuracy is difficult to assess, because pressure chamber measurements can only be compared with psychrometer methods, and when they do not agree one cannot know which is wrong. An indication of the accuracy is given by comparisons made by Boyer (1967a), Campbell and Campbell (1974), Baughn and

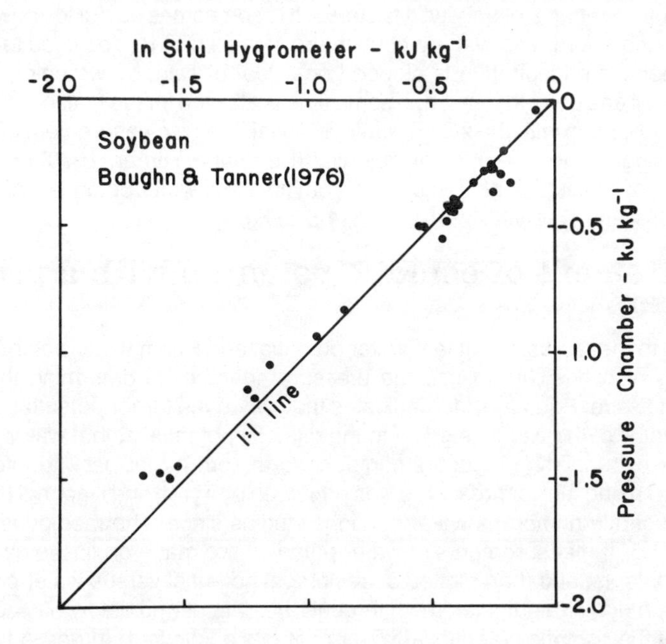

Figure 11.2 Comparison of pressure chamber and *in situ* hygrometer water potentials for soybean leaves, measured by Baughn and Tanner (1967b). Results were similar for potato, sunflower, pepper, and oat leaves.

Tanner (1976b), and others. The data from Baughn and Tanner are probably the best because the hygrometer measurements were made *in situ*, and the errors in the pressure chamber measurements were minimised due to careful attention to technique. Their results are shown in Fig. 11.2. Agreement between the two methods indicates that measurements with either should usually be within 150 J/kg of the other.

Biases exist in both measurements. The leaf hygrometer covers part of the leaf, stopping transpiration. This reduction in transpiration increases water potential in that part of the leaf (Brown and Tanner, 1981). The pressure chamber has a similar bias (Meiri *et al.*, 1975). Even though the pressure in the chamber is applied evenly to all the cells, those with the highest water potential start to contribute water to the xylem first when the tissue is pressurised. If gradients exist in the tissue when it is part of the intact plant, then the pressure chamber might give a reading that is weighted towards the potential of the wettest tissue. Since the errors in both measurements tend to make higher readings, the two measurements can be in good agreement, and both be in error (Brown and Tanner, 1981).

Measurement of xylem potential with a pressure chamber

Resistances to water transport within the plant are generally small except at the root endodermis, and somewhere in the vicinity of the leaf (Begg and Turner, 1970). The main body of xylem in the plant is therefore at a fairly constant potential, except for variations due to effects of gravity in tall trees. It is sometimes useful to know the water potential in the xylem. The leaf is a convenient probe which can be used to make that measurement. If transpiration is stopped on a leaf or twig, by wrapping the leaf in aluminium foil and polyethylene, and some time is allowed for the leaf to come to water potential equilibrium with the xylem water potential, then a measurement of leaf water potential will give the xylem water potential (Begg and Turner, 1970). Equilibration times of 1–2 h are sufficient. Typically, for daytime leaf water potentials of –1.5 to –2 kJ/kg, xylem water potentials will be –1 to –1.5 kJ/kg.

Measurement of osmotic potential with a pressure chamber

In addition to measurement of leaf water potential and xylem water potential of plant tissue, it is also possible to use the pressure chamber to determine the osmotic potential of tissue. Equation 11.2 indicates that, when the turgor potential is zero, the water potential of the tissue is equal to the osmotic potential at that water content. A method for extrapolating measurements made at zero turgor back to infer osmotic potentials at higher turgor pressures was proposed by Tyree and Hammel (1972), and has been used in numerous water relations studies since. The method is based on equation 11.5. If measurements of water potential are made on tissue over a range of water contents, and then plotted as reciprocal potential versus water content, the extrapolation of a straight line drawn through the data at and below $\psi_p = 0$ yields an estimate of the osmotic potential at water contents above zero turgor. A typical plot, with osmotic potential indicated, is shown in Fig. 11.3. Tyree and Hammel (1972) plotted reciprocal pressure as a function of volume of water extracted from the tissue to determine their osmotic potentials. This, and any other variable that is linearly

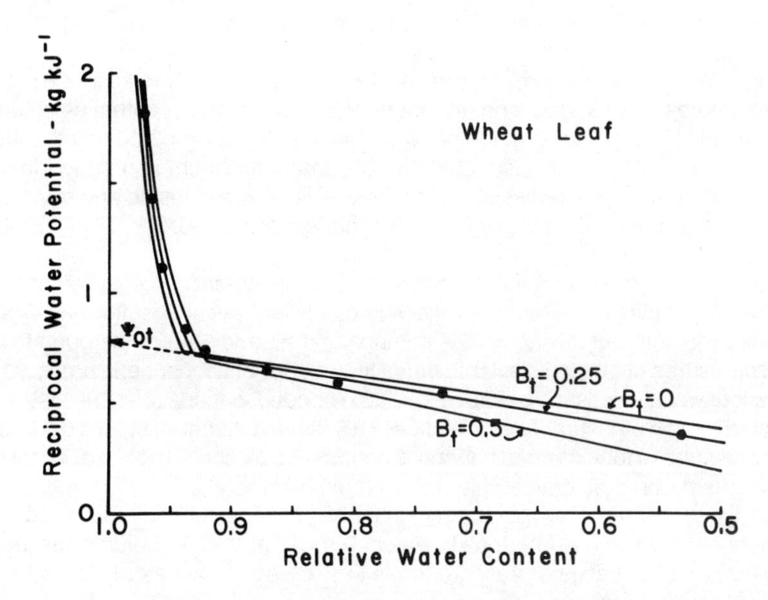

Figure 11.3 Pressure–volume curve for the wheat leaf in Fig. 11.1. Lines are drawn using equation 11.8 and the values in Fig. 11.1. Points are data for a wheat leaf. The extrapolation to $R = 1$ is used to find ψ_{ot}. Note that the value for ψ_{ot} obtained by the extrapolation is independent of B_t, as suggested by Tyree and Richter (1981).

related to the water content of the cells, will provide adequate data for determining osmotic potential.

Various methods have been used for hydrating and dehydrating the tissue to make osmotic potential measurements. The original method of Tyree and Hammel was to cut tissue, pressurise it, place the cut xylem under water, and then release the pressure. After a period of less than 2 h the tissue had imbibed enough water to be at full turgor, and dehydration could start. Successively higher pressures were set, and, at each pressure, the volume of water extracted at equilibrium was measured by a capillary attached to the stem.

Several modifications have been proposed for both the hydration and dehydration phases. Campbell *et al.* (1979) forced water into wheat plants under pressure to achieve rapid uptake and attainment of full turgor. Tyree *et al.* (1978) suggested, in addition to the above mentioned method, placing the stem or petiole in water overnight. In any case a starting potential of about -100 J/kg should be attained. Hellkvist *et al.* (1974) dehydrated the tissue by applying a set pressure to the tissue for a fixed time, and determined the change in water content by collecting the xylem fluid on absorbent material inside a small tube that fits over the stem. The tube was weighed after each pressure increment. Campbell *et al.* (1979) weighed leaves after each pressure increment, and allowed the leaves to dehydrate by evaporation rather than overpressure. Tyree *et al.* (1978) compared the Hellkvist method with their own, and indicated that the times used by Hellkvist *et al.* (1974) were too short for equilibration. They outline a method in detail which, they feel, gives accurate moisture release curves for leaves, and therefore accurate osmotic potential estimates. Comparisons have not been made between the evaporation method and the method of Tyree *et al.* (1978), but one would assume that equilibrium is reached faster by evaporation from the entire tissue surface than by overpressurising and removing all of the excess water through the xylem from all parts of the tissue.

The gas used to pressurise the chamber for osmotic potential measurements must be such that it does not affect the reflection coefficient of cell membranes. Tyree *et al.* (1978) recommend mixing 5-10% compressed air and 90-95% nitrogen to attain partial pressures of oxygen similar to those found in air at atmospheric pressure. This precaution was necessary because of the extended periods that the tissue was subjected to high pressure. Campbell *et al.* (1979) used nitrogen, but did not keep the tissue pressurised for extended periods. Compressed air should not be used for these measurements on a single sample, because cell membrane damage may result at high partial pressures of oxygen, and high carbon dioxide concentrations reduce the permeability of the cell membranes to water (Tyree *et al.*, 1978). Solutes can leak into the xylem, and the balance pressure tends to decrease with water loss, rather than increase.

Assumptions for osmotic potential measurement are similar to those for tissue water potential measurement, with additional assumptions that turgor pressure in the tissue is positive or zero, and that the solutes in the cells behave ideally (equation 11.5). As with the water potential measurement, no standard is available, and one can only compare various methods, each of which is subject to some error. Measurements using thermocouple psychrometers have most frequently been used as standards for comparison. Generally some difference exists between the two measurements. When such differences exist, the error is generally assumed to be an error in the psychrometer reading due to dilution of the cell solution by apoplastic water (Tyree, 1976). However, if negative turgor pressures in intact cells exist (Acock and Grange, 1981), then the pressure chamber measurements could be in error.

Measurement of matric potential with a pressure chamber

Boyer (1967b) used the pressure chamber for the measurement of matric potential in plant tissue. The measurement is made in a manner similar to that of water potential, but the tissue is frozen, before the measurement, to disrupt cell membranes. Since the walls of the xylem elements are permeable to solutes the osmotic component of the tissue fluid does not influence the reading, and at balance pressure the sum of the matric potential and the applied pressure is zero. The matric potential of the frozen tissue is therefore equal to the negative applied pressure. Once a reading is obtained, the tissue water content can be determined by weighing. Water can be removed from the tissue through evaporation or overpressurising the tissue in the pressure chamber, and a new balance pressure and water content can be determined. Fig. 11.4 shows a typical matric potential-water content characteristic for a frozen wheat leaf. Such curves are usually well described by equation 11.6 and interpolations and

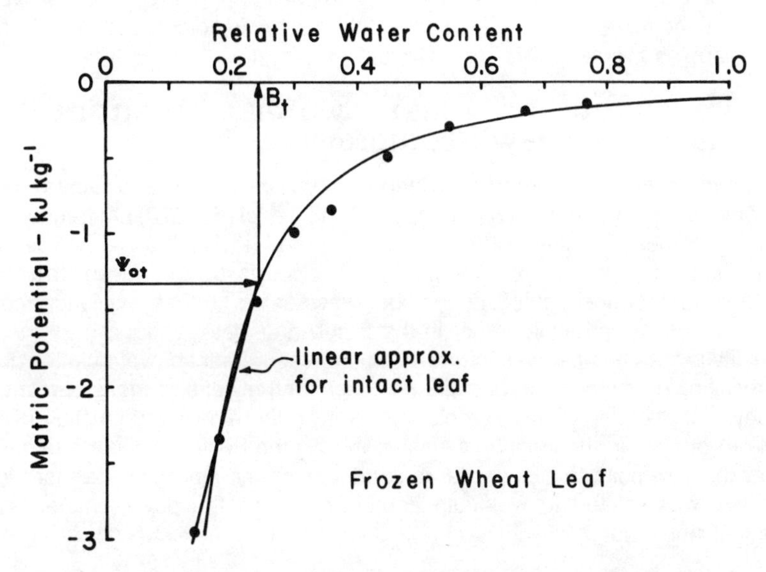

Figure 11.4 Matric potential–water content relation for a frozen wheat leaf. The smooth line is drawn using equation 11.6 with $b = 1.79$. Points are data for a wheat leaf. Matric potentials in the walls of intact cells are those below the ψ_{ot} line. The linear approximation used to obtain α in equation 11.8 for this portion of the curve is shown.

extrapolations are conveniently done by plotting the data on log-log graph paper or fitting equation 11.6 to the data.

The difficulty in the Boyer procedure comes in the interpretation of the data. It should be remembered that the data are for frozen tissue in which cell membranes have been ruptured and turgor is absent. We need to know how this potential relates to the matric potential which existed within the intact tissue. Water is held within and between cell walls and by the macromolecules within the symplast of the cell. At normal potentials the symplast is under positive hydrostatic pressure, so matric potential must be zero (Passioura, 1980). We therefore will focus on the matric potential of cell walls. Equation 11.9 indicates that the total potential in an intact cell wall is the sum of osmotic, matric and overburden components. The osmotic component is often negligible. The overburden component is some function of turgor pressure and arises from the pressure applied to the cell wall by the pressure inside the cell (equation 11.4) (Walter, 1963). The functional relation between $\psi_{\Omega w}$ and ψ_{pc} cannot be determined directly, but if we assume that the cell wall is sufficiently hydrated and sufficiently flexible that all of the turgor pressure is transmitted to the water in the wall ($\kappa = 1$) then, $\psi_{\Omega w} = \psi_{pc}$, and from equation 11.9, the matric potential in the cell wall is equal to the osmotic potential of the cell. A similar conclusion was reached by Walter (1963). Thus, for the matric potential curve in Fig. 11.4, only water contents below $\psi_m = \psi_{oc}$ have meaning (assuming that freezing does not alter the water holding properties of the matrix). With this assumption, Fig. 11.4 can be used to obtain an estimate of the apoplastic water fraction of intact leaves (Campbell *et al.*, 1979). The relative water content of the frozen leaf, corresponding to the osmotic potential of the cell, is taken as the apoplastic water fraction of the tissue.

Measurement of leaf water potential with thermocouple psychrometers

The thermocouple psychrometer (Spanner, 1951; Richards and Ogata, 1958) or hygrometer (Neumann and Thurtell, 1972; Campbell *et al.*, 1973) are frequently used for measurement of tissue water potential. Two modes of operation have been used. Spanner (1951), and numerous others since, excised tissue and placed it in a chamber with a small thermocouple. The chamber was sealed so that equilibrium could be established between liquid water in the tissue and water vapour in the chamber. Humidity in the chamber was inferred from a measurement of wet-bulb temperature depression. The small thermocouple was used for the wet bulb measurement, either by physically placing a small droplet of water on the measuring junction (Richards Method) or cooling the junction electrically using the Peltier effect to a temperature below the dew point to allow water to condense on the junction (Spanner Method). Humidity was related to wet bulb depression using the psychrometer equation (Monteith and Owen, 1958):

$$h=(e^{\circ}_{w}-\gamma^{*}(T-T_{w}))/e^{\circ} \qquad 11.14$$

where T is psychrometer temperature, T_w is wet bulb temperature, e° is the saturation vapour pressure at T, and e°_{W} is the saturation pressure at T_w. The apparent psychrometer constant, γ^* is usually determined from calibration with salt solutions having known osmotic potentials. Convenient sources of osmotic potential data are Lang (1967) and Campbell and Gardner (1971). Extensive tables of activity coefficients for calculating osmotic potentials are given by Robinson and Stokes (1965). Water potential is calculated from equation 11.14 using

$$\psi = (RT/M_w)\log_e(h) \qquad 11.15$$

or obtained by correlating psychrometer output with calibrating solution osmotic potential.

A second mode of operation, used by Neumann and Thurtell (1972), Campbell and Campbell (1974) and others, uses a small chamber attached directly to the intact plant. Again, water evaporates into the chamber to achieve equilibrium and the humidity in the chamber is measured with a fine thermocouple. Dew point temperature, rather than wet bulb is often measured with these instruments using special techniques and circuitry (Neumann and Thurtell, 1972) though a careful analysis of errors indicates little, if any advantage over the wet bulb method in many applications (Campbell *et al.*, 1973).

A modification of the Richards method, proposed by Boyer (1966) makes a measurement of thermocouple output with water on the measuring junction, and then another with a sucrose solution which has a potential approximately equal to that of the sample. Interpolation between the two readings, to a point where neither uptake nor loss occurs gives the water potential. The advantage of this isopiestic method, is that the effect of a finite diffusion resistance at the tissue surface is removed. However,

Boyer's conclusion that diffusion resistances cause significant errors is called into question by the fact that his isopiestic readings were drier than either the Richards or the Spanner measurements, which he made on the same leaf material. The Richards method might be expected to give somewhat wetter readings than the isopiestic method, because of the introduction of water into the chamber. The Spanner method, however, should give readings that, if anything, are slightly too dry because of the condensation of water out of the chamber onto the thermocouple (Monteith and Owen, 1958). Zollinger *et al.* (1966) made careful comparisons of leaf water potential measurements using the Richards and the Spanner methods and found no significant difference between the readings. It should be mentioned here that the dew point method of Neumann and Thurtell (1972) is also isopiestic.

Leaf water potential can be measured either *in situ* or on excised tissue samples using thermocouple psychrometers. The *in situ* psychrometer (or hygrometer, if dew point depression is measured) is a small cavity machined out of an aluminium or copper block. The cavity is sealed to the leaf and the aluminium or copper maintains isothermal conditions between the leaf and the cavity. Inside the cavity is a small thermocouple. Vapour from the leaf equilibrates with the cavity and the humidty in the cavity is measured with the thermocouple. Leaf psychrometers can be constructed according to the designs given by Neumann and Thurtell (1972) and Campbell and Campbell (1974) or purchased from commercial sources listed at the end of this chapter. The two main difficulties with the measurement result from non-isothermal conditions within the psychrometer and failure of the psychrometer cavity to reach vapour equilibrium with the leaf. The latter can be the result of incomplete sealing or high resistance to vapour transport between the leaf surface and the cavity. The high thermal conductivity and large thermal mass of the psychrometer body are intended to stabilise temperature within the chamber, but improved performance is obtained by enclosing the psychrometer in a thin layer of insulating material covered with a reflective covering (Brown and Tanner, 1981; Savage *et al.*, 1984). The insulation should cover as little of the leaf as possible, because reduced transpiration from the leaf will increase its water potential and result in measurement error.

Neumann and Thurtell (1972) used a small quantity of "Apiezon M" vacuum grease to seal their psychrometer to the leaf surface. Campbell and Campbell (1974) and Brown and Tanner (1981) used a mixture of wax and lanolin. A mixture should be obtained which is fairly hard at room temperature, but can be spread in a narrow ring around the outer edge of the psychrometer cavity before the psychrometer is placed on the leaf. The psychrometer should be pressed tightly enough to the leaf surface so that a good seal is obtained.

Maintaining low vapour diffusion between the leaf surface and the cavity is perhaps the most difficult and critical part of making accurate measurements with the leaf psychrometer. Neumann and Thurtell (1972) recommended using a small amount of xylene on a cloth and gently wiping the leaf surface. This appears to work well with maize, but kills the tissue of some other species. Brown and Tanner (1981) recommend abrading the cuticle with 600 grit carborundum. Savage *et al.* (1984) tried several abrasion treatments, and determined that a treatment similar to that of Brown and Tanner (1981), but with 400 grit carborundum gave best results. The abrasion treatment, if properly done, gives equilibrium within 15-30 minutes and does not result in errors due to damaged cells.

If the leaf is at temperature and water potential equilibrium with the chamber air, the errors in the *in situ* water potential measurement come only from the inherent psychrometer error (20-50 J/kg) and the fact that the psychrometer reduces the transpiration from the leaf and thus modifies the leaf water potential. The latter error is species dependent, but can be substantial in small-leaf species such as alfalfa (Brown and Tanner, 1981). In spite of these errors, agreement between *in situ* and pressure chamber measurements generally is good, as is indicated by Fig. 11.2. If the leaf is not at temperature equilibrium with the chamber air, large errors can result. A 0.001 K difference between the leaf and the chamber results in a 15 J/kg error in the measurement.

Psychrometric measurement of leaf water potential on excised samples is subject to other errors in addition to those for the *in situ* measurement. These errors result primarily from the changes that can take place in the tissue during the relatively long equilibrium times when the tissue is removed from its water source. Typically, the measurement is made by cutting strips or disks of tissue from leaves. These are placed in chambers of various designs, and allowed to equilibrate under constant temperature conditions for 3 to 6 hours. The vapour diffusion resistance of the leaf surface becomes large when the tissue is placed in the chamber, so that water coming from the low resistance cut surfaces of damaged cells can have a disproportionate influence on the humidity in the chamber (Barrs and Kramer, 1969). Talbot *et al.* (1975) recommend a ratio of length of cut cells to surface area less than 0.12 mm mm^{-2} for accurate measurements, though this must depend on species and cuticular resistance of the leaf surface. Measurements on small (6 mm diameter) disks, which are commonly used with one of the commercial psychrometers, have been shown to be in error because of this cut edge effect (Nelsen *et al.*, 1978).

An additional error results from changes in turgor which accompany growth when the tissue is separated from its water source. When an excised sample at high turgor is placed in a psychrometer chamber, cell enlargement can occur during the equilibrium process (Baughn and Tanner, 1976a). This decreases the turgor pressure and the leaf water potential. This error presumably would be small on mature tissue.

Another error results from respiratory heating of plant tissue in the psychrometer chamber. This error can be minimised by maintaining good thermal contact between the tissue and the chamber, and constructing the chamber from materials with high thermal conductivity.

As with the pressure chamber, large errors can result from water loss between the time of excision and the time the sample is sealed in the psychrometer chamber. Errors can also result from slow equilibration and from loss of internal water to hydrate external surfaces (Walter, 1963). Psychrometer chambers should have minimum volume, and be made from materials with minimum vapour adsorption. Polished metal surfaces are best. Leaf surfaces should be as clean as possible to minimise water vapour absorption. It would also appear helpful to try to reduce cuticular resistance, as is done with the leaf psychrometer, though I am not aware of any attempts at this.

The accuracy of the psychrometer with excised samples can only be assessed by comparison with other measurements. Baughn and Tanner (1976a) made careful comparisons of measurements on intact and excised samples using leaf psychrometers. Their results are shown in Fig. 11.5. As expected, readings tend to deviate from the 1:1 line at high turgor. It appears that with good technique, psychrometric measurements on excised tissue can be expected to be within 100 to 200 J/kg of the correct water potential at low water potential, with increased errors possible at high potential, particularly with the immature tissue. The magnitude of these errors would increase with time spent in the psychrometer chamber.

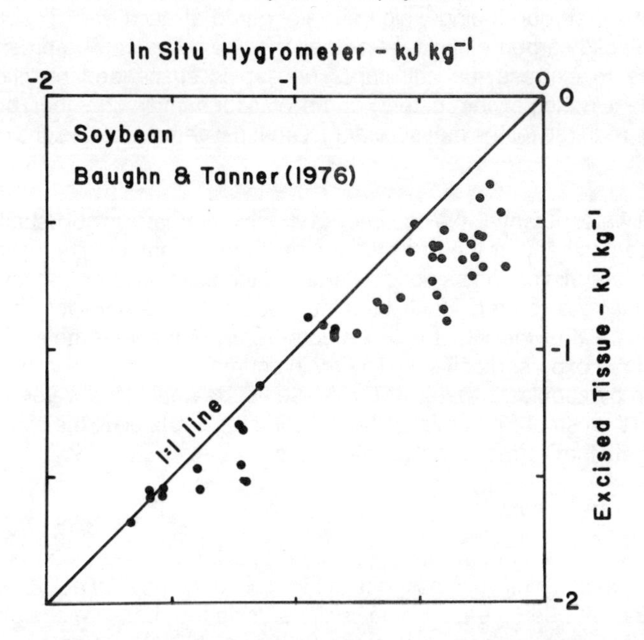

Figure 11.5 Comparison of *in situ* hygrometer and excised psychrometer water potentials for soybean leaves, measured by Baughn and Tannr (1967a). Results were similar for potato, sunflower, pepper and oat leaves, except that departures from the 1:1 line at high water potential are somewhat greater for soybean than for other species.

Measurement of osmotic potential with a thermocouple psychrometer

In order to measure osmotic potential with a thermocouple psychrometer, the cell membranes must be broken to release the turgor pressure in the cells. This is usually accomplished by quickly freezing the tissue with solid carbon dioxide or liquid nitrogen, but has also been accomplished with heat (Walter, 1963). Measurements can be made either on the frozen tissue or on the sap squeezed from the frozen tissue. The measurement on the frozen tissue is actually the sum of the matric and osmotic potentials of the frozen tissue (Warren Wilson, 1967) but the matric component is generally small at high relative water content (Fig. 11.4).

Walter (1963) warned that substantial errors are possible in the measurement of osmotic potential using frozen tissue because, when thawed, storage products are rapidly hydrolysed, lowering the osmotic potential. Brown and Tanner (1983) found that such errors ranged from 210 to 890 J/kg in alfalfa which was thawed for 3 hours. The heat treatment, proposed by Walter (1963), avoids this problem by denaturing the hydrolytic enzymes at the same time the cells are killed. If frozen tissue is used, the measurement should be made immediately following thawing.

A convenient method for measuring osmotic potential is to enclose the samples in short lengths of Tygon tubing which are stoppered at each end. The tubes can be frozen with solid carbon dioxide. After thawing, the tubes can be pressed between steel rollers to express the cell sap. The sap is suspended on filter paper for measurement in the psychrometer. For the Wescor sample chamber, only 2 mm^3 of solution are required for the measurement. Other psychrometer designs require larger quantities.

As with leaf water potential, the accuracy of the psychrometer is not usually the limiting factor in the accuracy of the osmotic potential measurement. The psychrometer is capable of determining the osmotic potential of the solution placed in the chamber to within ± 20 J/kg or better. Much larger uncertainties are invovled in inferring the osmotic potential of the intact cell contents from the measurement of the osmotic potential of the expresed cell sap. The main source of error is usually the dilution of the cell sap by apoplastic water. If B represents the apoplastic water fraction in the tissue, then the osmotic potential of the cell contents is related to the osmotic potential measured with the thermocouple psychrometer by

$$\psi_{oc} = \psi_{psych}/(1-B) \quad 11.16$$

Apoplastic water fractions ranging from 0.05 to 0.3 have been inferred from matric potential characteristics such as those in Fig. 11.4 (Campbell *et al.*, 1979). Values as high as 0.5 have been measured on conifers using the pressure chamber (Hellkvist *et al.*, 1974). Both methods involve assumptions which could cast doubt on these numbers. However, in spite of these doubts it is almost certain that most plant tissues have a measurable apoplastic water fraction. The osmotic potential for this apoplastic water is generally assumed to be near zero. Corrections of psychrometer measurements for apoplastic water are made using equation 11.16 and an estimate of the apoplastic water fraction. The apoplastic water fraction can be estimated from Fig. 11.4, using a matric potential characteristic for the tissue, or by assuming negative turgor in cells is negligible and making a psychrometer or pressure chamber measurement on obviously flaccid tissue so that $\psi = \psi_o$. The tissue is then frozen to measure ψ_{psych}. Equation 11.16 can be solved for B. It should be possible to determine a mean value of B for the species and the environment of interest, and then use this to correct all of the psychrometrically measured osmotic potentials. The apoplastic water fraction has some dependence on tissue water potential, but, as can be seen from Fig. 11.4, this is small enough to be neglected over the range of interest of most plant water studies.

Measurement of low water potentials

The typical range of the thermocouple psychrometer using Peltier cooling is 0 to –5 or –6 kJ/kg. Wilson and Harris (1968) were able to extend this range by condensing water on the thermocouple over a dilute salt solution and then allowing the water to evaporate over the dry sample. The method was extended by Campbell and Wilson (1972) and Wiebe (1981) and has been used for measuring the water potentials of dry seeds, foods, and other substrates. Calibration is generally accomplished using salt solutions (Greenspan, 1977). The method has been shown to work well with either multiple sample chambers (Campbell *et al.*, 1966) or with the Wescor single sample units as modified by Wiebe (1981). This method, while useful, is time consuming (5 to 10 minutes for water condensation per sample) and the amount of water which can be held on the thermocouple is so small that the reading is held for only a short time at low water potentials, allowing insufficient time for equilibration with the sample and accurate reading.

In an effort to increase the amount of water held on the junction, we tried a modification of the Richards method (Richards and Ogata, 1958). Instead of using the silver loop, the thermocouple junction was covered with a 0.5 mm diameter ceramic ball. This thermocouple was mounted in a sample changer modified from that of Campbell *et al.* (1966) so that it has 10 sample cups and is surrounded by a massive aluminium heat sink to eliminate the need for a constant temperature bath. The thermocouple can be moved into a sealed position over each of the 10 sample cups. One of the sample locations is set up with a small well filled with distilled water. The container with the water is raised to dip the ceramic coated thermocouple in the water. The sample selector is then turned to bring the thermocouple over the unknown sample, and the wet bulb depression is measured. This method of reading dry samples has extended the measurement range so that reliable measurements can be made at sample humidities below 0.1 (water potentials below –300 kJ/kg). The accuracy of such measurements, inferred from calibrations with saturated salt solutions, is better than 0.005, which is about the uncertainty in the standards (Greenspan, 1977).

Direct measurement of turgor pressure

Turgor can be inferred from measurement of total and osmotic potential using equation 11.2. However, rapid non-destructive measurements are also possible. Heathcote *et al.* (1979) measured the deformation of a circular area of leaf lamina resulting from an applied stress. The deformation was related to the bulk elastic modulus of the leaf, which, in turn, was linearly related to the turgor pressure. Unfortunately, the relation between turgor and elastic modulus differs from species to species, so a calibration is necessary for each species. Calibration is accomplished by measuring ψ_o and ψ_l as previously described.

Future developments and needs

There is an immediate need to determine the extent to which negative turgor exists in cells. If sizeable negative turgor pressures are possible, then apoplastic water fractions are likely to be much smaller than are measured with the pressure chamber (Tyree, 1976) or inferred from equation 11.16.

Looking farther into the future, we need to develop instrumentation for characterising the spatial and temporal variation in plant water status in more detail. This would require methods which are faster and, hopefully, simpler than those presently available. Remote sensing, using effects of water stress on optical properties of leaves and canopies should be developed. Scope for such development is indicated by the fact that farmers, and other trained observers can often 'see' stress in a crop when scientists have difficulty measuring it with the instruments described in this chapter.

Finally, there is a continuing need to relate plant response to measurements of plant water status. The plant response is the integral, over time, of some response function multiplied by an appropriate potential. Before we can relate "water stress" to any of the measurements described in this chapter, we need to have a better understanding of the response function and its relevant potential.

Manufactures and Suppliers

Pressure chambers

Chas. W. Cook and Sons, Ltd. Perry Bar, Birmingham, England.

P.M.S. Instrument Co., 2750 NW Royal Oaks Drive, Corvallis, Oregon 97330, U.S.A.

Professor P.G. Jarvis, Department of Forestry and Natural Resources, University of Edinburgh, Darwin Building, King's Buildings, Mayfield Road, Edinburgh EH9 3JU, Scotland.

Soil Moisture Equipment Co., P.O. Box 30025, Santa Barbara, CA 93105, U.S.A.

Psychrometers and associated electronics

Campbell Scientific, Inc., Box 551, Logan, Utah 84321, U.S.A.

Decagon Devices, Inc., N.W. 800 Fisk, Pullman, WA 99163, U.S.A.

J.R.D. Merrill Specialty Equipment, RFD Box 140A, Logan, Utah 84321, U.S.A.

Wescor, Inc., 459 South Main, Logan, Utah 84321, U.S.A.

References

Acock, B. & Grange, R.I. (1981). Equilibrium models of leaf water relations. *In* Mathematics and Plant Physiology, eds. D.A. Rose and D.A. Charles-Edwards. London: Academic Press.

Andrews, F.C. (1976). Colligative Properties of simple solutions. Science, N.Y., **194**, 567-571.

Aslyng, H.C. (1963). Soil physics terminology. Int. Soc. Soil Sci. Bull. 2317.

Barrs, H.D. & Kramer, P.J. (1969). Water potential increase in sliced leaf tissue as a cause of error in vapour phase determinations of water potential. Pl. Physiol., **44**, 959-964.

Baughn, J.W. & Tanner, C.B. (1976a). Excision effects on leaf water potential of five herbaceous species. Crop Sci., **16**, 184-190.

Baughn, J.W. & Tanner, C.B. (1976b). Leaf water potential: comparison of pressure chamber and *in situ* hygrometer on five herbaceous species. Crop Sci., **16**, 181-184.

Begg, J.E. & Turner, N.C. (1970). Water potential gradients in field tobacco. Pl. Physiol., **46**, 343-346.

Boyer, J.S. (1966). Isopiestic technique: measurement of accurate leaf water potentials. Science, N.Y., **154**, 1459-1460.

Boyer, J.S. (1967a). Leaf water potentials measured with a pressure chamber. Pl. Physiol., **42**, 133-137.

Boyer, J.S. (1967b). Matric potentials of leaves. Pl. Physiol., **42**, 213-217.

Briggs, L.J. (1950). Limiting negative pressure of water. J. appl. Phys., **21**, 721-722.

Brown, P.W. & Tanner, C.B. (1981). Alfalfa water potential measurement: a comparison of the pressure chamber and leaf dew-point hygrometer. Crop Sci., **21**, 240-244.

Brown, P.W. & Tanner, C.B. (1983). Alfalfa osmotic potential: a comparison of water-release curve and frozen-tissue methods. Agron. J., **75**, 91-93.

Campbell, G.S. & Campbell, M.D. (1974). Evaluation of a thermocouple hygrometer for measuring leaf water potential *in situ*. Agron. J., **66**, 24-27.

Campbell, G.S. & Gardner, W.H. (1971). Psychrometric measurement of soil water potential: temperature and bulk density effects. Proc. Soil Sci. Soc. Am., **35**, 8-12

Campbell, G.S. & van Schilfgaarde, J. (1981). Use of SI units in soil physics. J. Agron. Educ., **10**, 73-74.

Campbell, G.S. & Wilson, A.M. (1972). Water potential measurements of soil samples. *In* Psychrometry in Water Relations Research, eds. R.W. Brown and B.P. van Haveren, Utah Agric. Exp. Sta., Logan. pp. 142-148.

Campbell, E.C., Campbell, G.S. & Barlow, W.K. (1973). A dewpoint hygrometer for water potential measurement. Agric. Meteorol., **12**, 113-121.

Campbell, G.S., Zollinger, W.D. & Taylor, S.A. (1966). Sample changer for thermocouple psychrometers: Construction and some applications. Agron. J., **58**, 315-318.

Campbell, G.S., Papendick, R.I., Rabie, E. & Shayo-Ngowi, A.J. (1979). A comparison of osmotic potential, elastic modulus, and apoplastic water in leaves of dryland winter wheat. Agron. J., **71**, 31-36.

Gandar, P.W. & Tanner, C.B. (1975). Comparison of methods for measuring leaf and tuber water potentials in potatoes. Am. Potato J. **52**, 387-397.

Greenspan, L. (1977). Humidity fixed points of binary saturated aqueous solutions. J. Res. natn. Bur. Stand., **81A**, 89-96.

Heathcote, D.G., Etherington, J.R. & Woodward, F.I. (1979). An instrument for non-destructive measurement of the pressure potential (turgor) of leaf cells. J. exp. Bot., **30**, 811-816.

Hellkvist, J., Richards, G.P. & Jarvis, P.G. (1974). Vertical gradients of water potential and tissue water relations in Sitka spruce trees measured with the pressure chamber. J. appl. Ecol., **11**, 637-668.

Hillel, D. (1980). Fundamentals of Soil Physics. New York: Academic Press.

Lang, A.R.G. (1967). Osmotic coefficients and water potentials of sodium chloride solutions from 0 to 40°C. Aust. J. Chem., **20**, 2107-2023.

Leach, J.E., Woodhead, T. & Day, W. (1982). Bias in pressure chamber measurements of leaf water potential. Agric. Meteorol., **27**, 257-263.

Meiri, A., Plaut, Z. & Shimshi, D. (1975). The use of the pressure chamber technique for measurement of the water potential of transpiring plant organs. Physiologia Pl., **35**, 72-76.

Monteith, J.L. & Owen, P.C. (1958). A thermocouple method for measuring relative humidity in the range 95-100%. J. scient. Instrum., **35**, 443-446.

Myrold, D.D., Elliott, L.F., Papendick, R.I. & Campbell, G.S. (1981). Water potential-water content characteristics of wheat straw. J. Soil Sci. Soc. Am., **45**, 329-333.

Nelsen, C.E., Safir, F.R. & Hanson, A.D. (1978). Water potential in excised leaf tissue: comparison of a commercial dewpoint hygrometer and thermocouple psychrometer of soybean, wheat and barley. Pl. Physiol., **61**, 131-133.

Neumann, H.H. & Thurtell, G.W. (1972). Peltier cooled thermocouple dewpoint hygrometer for *in situ* measurement of water potential. *In* Psychrometry in Water Relations Research, eds. R.W. Brown and B.P. van Haveren, Utah Ag. Exp. Sta., Logan, pp. 103-112.

Passioura, J.B. (1980). The meaning of matric potential. J. exp. Bot., **31**, 1161-1169.

Puritch, G.S. & Turner, J.A. (1973). Effects of pressure increase and release on temperature within a pressure chamber used to estimate plant water potential. J. exp. Bot., **24**, 342-348.

Richards, L.A. & Ogata, G. (1958). Thermocouple for vapour-pressure measurement in biological and soil systems at high humidity. Science, N.Y., **128**, 1089-1090.

Ritchie, G.A. & Hinckley, T.M. (1975). The pressure chamber as an instrument for ecological research. Adv. Ecol. Res., **9**, 165-254.

Robinson, R.A. & Stokes, R.H. (1965). Electrolyte Solutions. London: Butterworths.

Savage, M.J., Wiebe, H.H. & Cass, A. (1984). Effect of cuticular abrasion on thermocouple psychrometer *in situ* measurement of leaf water potential. J. exp. Bot., **35**, 36-42.

Slatyer, R.O. (1967). Plant-Water Relationships. New York: Academic Press.

Spanner, D.C. (1951). The Peltier effect and its use in the measurement of suction pressure. J. exp. Bot., **2**, 145-168.

Spanner, D.C. (1973). The components of the water potential in plants and soils. J. exp. Bot., **24**, 816-819.

Talbot, A.J.B., Tyree, M.T. & Dainty, J. (1975). Some notes concerning the measurement of water potentials of leaf tissue with specific reference to *Tsuga canadensis* and *Picea abies*. Can. J. Bot., **53**, 784-788.

Taylor, S.A. & Ashcroft, G.L. (1972). Physical Edaphology. San Francisco: W.H. Freeman.

Turner, N.C. & Long, M.J. (1980). Errors arising from rapid water loss in the measurement of leaf water potential by the pressure chamber technique. Aust. J. Plant Physiol., **7**, 527-537.

Tyree, M.T. (1976). Negative turgor in plant cells: fact or fallacy. Can. J. Bot., **54**, 2738-2746.

Tyree, M.T. & Hammel, H.T. (1972). The measurement of the turgor pressure and the water relations of plants by the pressure-bomb technique. J. exp. Bot., **23**, 267-282.

Tyree, M.T. & Richter, H. (1981). Alternative methods of analysing water potential isotherms: some cautions and clarifications. J. exp. Bot., **32**, 643-653.

Tyree, M.T., MacGregor, M.E., Petrov, A. & Upeniaks, M.I. (1978). A comparison of systematic errors between the Richards and Hammel methods of measuring tissue-water relations parameters. Can. J. Bot., **56**, 2153-2161.

Walter, H. (1963). Zur Klarung des spezifischen Wasserzustandes in Plasma und in der Zellwand bei hoheren Pflanzen und seine Bestimmung. Ber. dt. bot. Ges., **76**, 40-53. (English translation in Adv. Front. Plant Sci., **14**, 173-218).

Warren Wilson, J. (1967). The components of leaf water potential. Aust. J. biol. Sci., **20**, 329-367.

Wiebe, H.H. (1981). Measuring water potential (activity) from free water to oven dryness. Pl. Physiol., **68**, 1218-1221.

Wilson, A.M. & Harris, G.A. (1968). Phosphorylation in crested wheat grass at low water potentials. Pl. Physiol., **43**, 61-65.

Zollinger W.D., Campbell, G.S. & Taylor, S.A. (1966). A comparison of water potential measurements made using two types of thermocouple psychrometer. Soil Sci., **102**, 231-239.

Chapter 12: Measuring plant growth and structure

P. V. Biscoe and K. W. Jaggard
Broom's Barn Experimental Station, Higham,
Bury St. Edmunds, Suffolk, IP28 6NP, England.

Introduction

Interest in measurement of plant growth arises from a desire to understand the extent to which alterations to or variations in the environment, either above or below ground, will affect the growth of a species. In some instances these changes may be entirely natural, for example the growth of a species in different seasons, but in others it may be artificial, for example when different treatments are applied in the same season. It is also important to know how these are caused, either by directly affecting the size and/or the efficiency of the photosynthetic system for producing dry matter or by altering its distribution between different plant organs. These desires prompted Blackman (1919), Briggs *et al.* (1920), Watson (1952) and others to develop the various concepts of the "classical" growth analysis approach for examining plant performance. These workers only had access to simple measuring techniques and were able, often laboriously, to determine the dry weight and size of plant organs, particularly those above ground. Using these determinations, plant dry weight and area of leaves, they developed systems for expressing the rate of plant growth in relation to time, to the amount of dry matter present at the start of the measurement period and to the size of the carbon-assimilating system. Many of these approaches had considerable drawbacks, and the concepts and associated difficulties of growth analysis have previously been reviewed in detail (cf. Kvet *et al.*, 1971; Evans, 1972). In attempting to overcome some of these difficulties, mathematical treatments of the data have become increasingly complex. Sophisticated curve-fitting procedures (Causton and Venus, 1981) have been used which, at their simplest, were likely to obscure real variations caused by fluctuations in the environment. The more complex procedures (Hunt, 1982) can describe growth accurately but, like all detailed analyses, they are of limited value unless the original data are very reliable.

Despite these advances many difficult problems remained. The most intractable of these were:

(a) the size of the sampling errors associated with making destructive measurements, and

(b) the measure of the efficiency of use of environmental resources by plants, the net assimilation rate (NAR), defined as the weight of dry matter produced per unit leaf area per unit time, was not independent of the size of the photosynthetic system.

However, the need for effective and reliable measurements of plant growth is greater than ever. Earlier chapters of this book have demonstrated many significant advances in the measurement of the plant environment, but these will be of limited value unless the plant's responses to its environment can be measured with equivalent accuracy and precision. This chapter will highlight some of the recent developments in measuring plant growth that have been made in an attempt to overcome some of the problems encountered in the earlier work and draw attention to existing difficulties. Although the authors' training and experience are in agricultural science and many of the examples are selected from this discipline, the principles apply to other circumstances in which plant growth has to be measured.

Sampling and errors

The advantages of destructive measurements of plant growth are their simple requirements in terms of equipment and technical skill, although it is very labour intensive to harvest and process even small samples of plants. Their chief disadvantage is the large sampling errors associated with harvesting relatively few plants. Natural variations in plant size often determine that many are needed to reflect accurately the average performance of the population. Also, when a small group of plants is harvested there are often large errors in assessing the area occupied by those plants. The large errors associated with destructively harvesting few plants has had two major consequences. First, the sample size and number of replicates has to be as large as possible to reduce the errors. Consequently the area of suitable plant materials must be increased, which can present problems either when natural communities are being sampled or of ensuring that treatments are applied uniformly, e.g. the infection of plants with specific diseases. This increases the labour and time required to complete the measurements, and therefore tends to reduce the frequency of harvests. Second, even in very accurate sampling programmes the errors could be such that the standard deviation of a population of samples is 10% of the sample mean, hence in absolute terms the errors increase as the plants grow. All too frequently these aspects are not given sufficient consideration and sampling programmes throughout the life of the crops are undertaken when the anticipated differences between seasons or treatments are far too small to be detected reliably.

These points are illustrated by comparing the growth of sugar beet and winter wheat crops in the northern hemisphere. Sugar beet, unlike wheat, is out-breeding so that a variety contains much genetic variation, also the population density is low so that a sample area contains relatively few plants in comparison with wheat and most other seed-bearing or forage crops. Beet emerges at the end of April and by early September the dry matter yield might be about 1500 g m^{-2}. At this time a good, healthy crop should have a growth rate of approximately 100 g m^{-2} (ground area) wk^{-1}, which consistently declines throughout the rest of the season until harvest in October, November or December. In contrast, the dry matter yield of winter-sown wheat at the beginning of April is about 200 g m^{-2} and the crop grows rapidly, at an approximately constant rate of 200 g m^{-2} wk^{-1} from then until the end of June when over 90% of the final dry matter yield has been produced. In an experiment with 6 replicates and a standard deviation of 10% of the mean, the standard error of the difference between the mean yield at two consecutive harvests halfway through the growth of a sugar beet crop might be 85 g m^{-2} compared with only 55 g m^{-2} for wheat. For sugar beet, therefore, a two week interval between harvests in September will occasionally give rise to apparently negative rates of growth and this will occur more frequently as the season progresses. This interval between harvests therefore provides very little reliable information about the rate of crop growth, but during longer periods environmental variations are likely to be so large that almost any interpretation can be applied to the data. This problem becomes more acute as the season progresses; the yield and hence absolute size of the sampling errors increases while the potential weight increment between harvests decreases. In contrast, throughout most of its life, wheat has small sampling errors relative to its growth rate, so making a programme of destructive sampling a realistic proposition for investigating the effect of environmental factors on crop growth.

Data interpretation

As previously stated, in conventional growth analysis a major limitation to data interpretation is that the measure of the photosynthetic efficiency of the leaf canopy, NAR, is very dependent upon, and negatively correlated with, the area of leaf present on the plant (Watson, 1958). Therefore, any environmental factor or treatment which alters the amount of leaf will change NAR, irrespective of any effect it has either on the interception of sunlight or the efficiency with which the canopy converts light to plant dry weight. Recently, Monteith (1977) has demonstrated that this serious restriction can be overcome by expressing the rate of plant growth in terms of the amount of solar radiation, S, which is intercepted and the net conversion coefficient of intercepted radiation into dry matter, ϵ. The relation may be expressed in the form,

$$dW/dt = \epsilon f S \qquad 12.1$$

where f is the fraction of the incident solar radiation intercepted by the crop. This approach provides a very powerful growth analytical technique because it isolates the effects of the major environmental factor influencing growth, radiation, and allows the effects of other environmental variables, or treatments, to be examined critically in

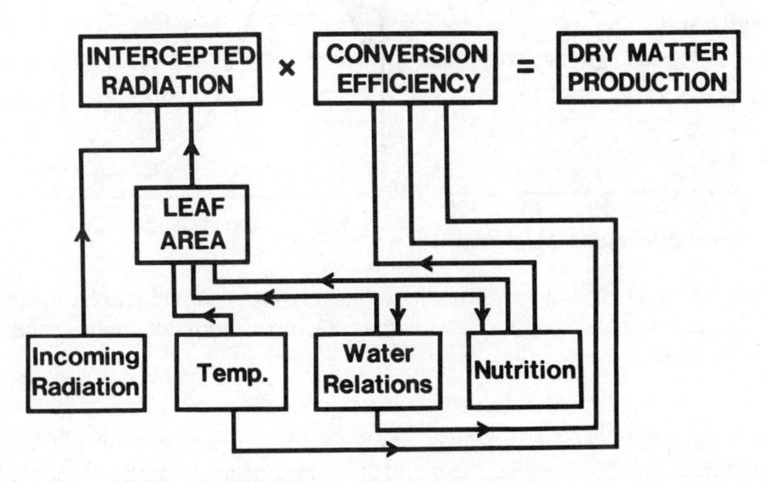

Figure 12.1 A diagram representing possible influences of environmental variables on the growth analysis system proposed by Monteith (1977).

terms of either the amount of radiation intercepted or the conversion coefficient to dry matter (Fig. 12.1). This form of analysis has been used to investigate the possible role of a plant growth regulator (PGR) for sugar beet crops, which laboratory studies had shown to decrease leaf area but increase both leaf thickness and the rate of photosynthesis per unit leaf area (Jaggard *et al.*, 1982). The aim of the experiment was to determine the effect of the changes in leaf structure on photosynthesis by the canopy. Fig. 12.2a shows that the rate of dry matter production per unit of solar radiation intercepted was unchanged by the growth regulator treatment and that any differences in dry matter yield were the result of changes in the amount of radiation intercepted. However, an analysis based on NAR (Fig. 12.2b) indicates that the PGR treatment made the canopy more efficient, but as the season progressed the efficiency of both canopies declined. The former indication resulted from the PGR reducing leaf area and mutual shading, but maintaining radiation interception, so that both treated and control crops produced similar amounts of dry matter. Thus the PGR-treated crop had a larger net assimilation rate. The decline in the net assimilation rates as the season progressed was caused by crops having relatively constant leaf area and intercepting a similar proportion of the declining amount of incident radiation throughout late summer and autumn. Therefore, during this period the plant dry weight increments declined in relation to leaf area, resulting in progressively lower NAR's. This dependence of NAR on leaf area and radiation is well known (Kvet *et al.*, 1971) and makes this measure of canopy performance an inappropriate discriminator between these treatments.

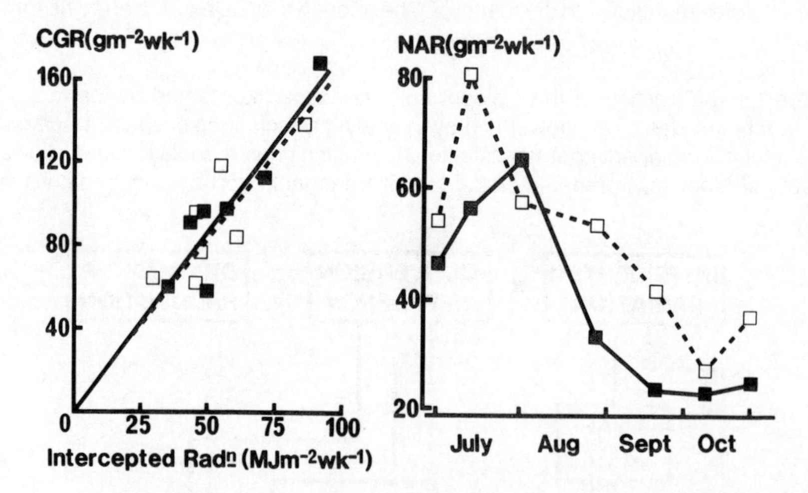

Figure 12.2 A contrast between two methods of assessing the photosynthetic activity of two sugar beet canopies; □, growth regulator treated and, ■, control.

In either a classical growth analysis study or the approach outlined above measurements of growth of whole plants and of some constituent parts, e.g. leaf area index (LAI), are required and these are now considered. However, this chapter makes no attempt to review all of the methods for making and interpreting measurements of plant growth and structure because, in most instances, recent comprehensive reviews exist and these are cited where appropriate. Instead, we attempt to direct attention to some techniques which are relatively simple and quick to use; while they can be applied to classical growth analysis they are more appropriate for the approach proposed by Monteith in 1977. We have done this deliberately because his proposals are relatively recent and although, as yet, they have not been widely used or tested they promise a very powerful technique for scientists who wish to analyse plant growth in response to the environment.

Measurements of plant communities

Destructive

The techniques for sampling plants are well established (e.g. Hudson, 1939; Evans, 1972). Similarly, accurate balances for the measurement of plant weight have been available for over a century (Lawes, 1850). Recent developments have made balances cheaper and much more robust, without loss of accuracy, and enabled the output to be recorded and processed directly using modern microprocessors and computers (McNicol *et al.*, 1982).

Measurements of the dimensions, area and volume of various plant organs are more time consuming and often the shapes of the structures are quite complex. Therefore researchers have directed considerable attention to devising rapid and more or less accurate assessment systems. Much attention has been given to the measurement of leaf area because leaves are important as the site of photosynthesis, also leaf area is an important parameter in both classical growth analysis and in the approach proposed by Monteith (1977) because it enables the proportion of the incident radiation being intercepted by the crop to be estimated (Szeicz, 1974). Many of the measurement systems were critically reviewed by Kvet and Marshall (1971), but some valuable techniques have been developed since then.

Automatic photoelectric planimeters which used flexible polythene belts to feed the leaf material past a scanning head at constant speed are accurate, rapid and very useful for work with species whose leaves are narrow, e.g. grasses. Wider leaves have to be cut into strips to fit the machine; this is tedious and stains the belts with sap and chlorophyll so that they have to be cleaned regularly to avoid spurious results.

We have found that the area of large, wide leaves like those of sugar beet and many *Brassica* species are better measured by a photographic technique using high-contrast, fine grain black and white film. To achieve sharp images and high contrast we have found line copying film, e.g. Kodak Ortholith 3, to be very satisfactory. The leaves to be measured are stripped off the plant and laid on a matt, white plastic, laminate surface, which has the advantage of being easy to clean and giving good contrast without highlights. To eliminate any shadows with sugar beet leaves it is easier if the leaves ar allowed to wilt slightly beforehand. Once developed, the photographic negatives are used as the image for a television camera which transfers the picture to the electronic measuring grid of a Quantimet 720 [1] image analysing computer (Fisher, 1971). Briefly, this instrument divides the measurement area into a grid of 5 10^5 "picture-points" and the image to be analysed is superimposed upon this. At its simplest the instrument can be set to determine the number of "picture-points" in light or dark portions of the image and can then calculate, by reference to a standard dimension on the photograph, the area of the leaves. It can also measure the number of leaves, together with the perimeter or other critical dimensions of individual leaves.

This system for measuring leaf area is potentially very accurate, provided that the camera used to photograph the leaves is fitted with a lens which is free from serious distortions. However, care should be taken to fill the film frame with the image of the leaves so that as many "picture-points" as possible are covered by the image of the leaf on the computer screen. This is particularly important if long, narrow leaves like grasses are being measured because then, a very small image on the Quantimet screen can be the subject of large edge-effect errors. Similar problems occur if leaf area is assessed by counting the number of squares covered by a leaf on a graph paper grid – the method is very inaccurate if the grid size is too large in relation to the leaf area. Other sources of error result from the leaves not lying flat and thus casting shadows which appear on the negative either as leaf if the film is over exposed or as intermediate tone. In the latter case the machine has to discriminate between shadow and leaf; this can be done relatively easily and accurately but it takes time. Having the photographic area well lit, if necessary from below so that leaves can be flattened with glass or transparent plastic, may be worthwhile in some circumstances. This method of measuring leaf area has the advantage that it provides a permanent image of the leaves, which can be useful where a record is needed of their size and shape in relation to their order of appearance on the plant.

Not everyone has access to expensive image-analysing computers, and much cheaper alternatives but with less, although very acceptable, resolution are the systems which depend upon integrating the time taken to scan a high-contrast image on a television monitor (Lawrence, 1969) [2]. The system accuracy again depends upon having the leaves flat and upon the quality of the television camera, but there is no permanent record of the leaves.

Non-destructive

Previously, in a classical growth analysis study, the area of the assimilating surface had to be measured to enable the full range of calculations to be done. Using the approach described by Monteith (1977) it is unnecessary to make physical measurements of leaf area, which saves much tedious work. Instead, the amount of light intercepted by the canopy has to be measured, which is relatively straightforward now that suitable solarimeters are available (Sheehy, Chapter 2) whose output can be conveniently integrated over time. However, the comparison of even a few treatments is expensive, especially in plant stands with wide or large irregular spaces between plants where many tube solarimeters have to be used in concert to give representative readings. If permanently positioned, these can make plant communities difficult to manage without risking breakages. Alternative non-destructive methods for assessing radiation interception, such as "fish-eye" photographs and inclined point quadrat estimates of canopy closure, have been developed but either the interpretation of the photographs or the collection of the data is tedious (see Kvet and Marshall, 1971). Also, the bulk of the camera and the height of the lens above the ground makes the "fish-eye" technique of photographing the canopy from below unsuitable for dense or low growing crops (Anderson, 1971).

Figure 12.3 A negative on infra-red sensitive film typical of those used to measure ground cover by sugar beet canopies.

We have developed a simple method for estimating the proportion of land area covered by foliage that could readily be interpreted and used quantitatively in analyses of crop growth. The technique uses the differential reflection of infra-red radiation from leaves and the soil to produce a high-contrast photographic negative (Fig. 12.3) from which the proportion of land covered by leaves can be measured either using image analysis or by projecting the negatives on to a suitably marked grid of dots and counting the points in light and dark areas (Henicke, 1963). With sugar beet crops, an area 1.5 x 1.0 m has been photographed from 1.5 to 2.0 m vertically above the top of the canopy using a lens of focal length 28 mm, a Kodak Wratten 88 filter and a high-speed infra-red sensitive film, e.g. Kodak 2481. Strong contrast between the images of the leaves and soil was obtained in diffuse sunlight; in bright sunshine the subject had to be shaded with a loosely-woven polypropylene screen. The area to be photographed and hence the camera height may need to be altered in other crops and, as with tube solarimeters, when working with row crops care is needed to ensure the space between the rows is accurately represented. The area being photographed needs careful preparation to remove large stones from the soil surface and unwanted vegetation, e.g. weeds in an agricultural crop, because these will appear dark, like leaves, on the negative and so cause errors in the calculation of ground cover.

Results from analyses using this technique indicate that the percentage of solar radiation intercepted is directly proportional to ground cover for most of the growing season of crops in temperate latitudes (Fig. 12.4). However, for sugar beet during October and November in England a hysteresis has been observed when leaf area and the canopy structure alters, due to senescence of the larger leaves but nevertheless radiation interception is maintained. Whether this hysteresis is caused by changes in the mean orientation of the intercepting organs or by changes in the solar altitude is being investigated.

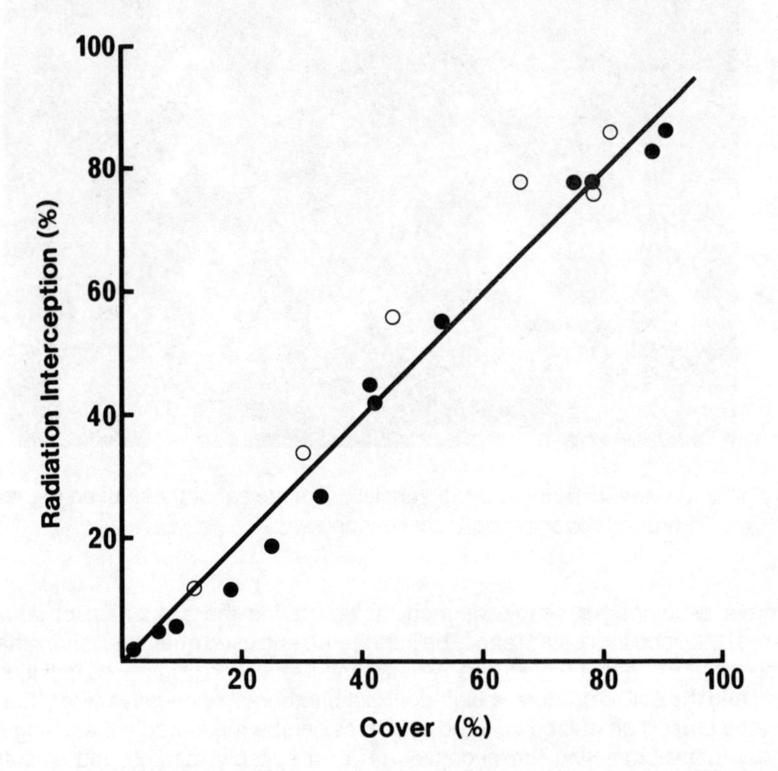

Figure 12.4 The relation between ground cover and radiation interception in sugar beet.

One of the most exciting prospects for non-destructive assessments of crop growth is to use reflection characteristics of the canopy and to sense these remotely. Various researchers around the world are attempting to do this using satellite imagery like that prouced by Landsat (Cipra *et al.*, 1980; Wiegand *et al.*, 1979). However, in Britain the use of satellites for measuring the quality of reflected light is very restricted by the frequent cloud cover, as the satellite pictures on televised weather forecasts clearly demonstrate. Measurements taken from low-flying aircraft offer more scope.

Kumar (1981) and Kumar and Monteith (1982) have shown theoretically that in equation 12.1 the fraction, *f*, of incident light which is intercepted by foliage, can be easily estimated from the ratio of infra-red to red radiation reflected from the land surface. This technique makes use of the strong absorption of red and reflection of infra-red radiation by green foliage. The use of a spectral ratio also eliminates many of the problems associated with variation in incident irradiance while the measurements are being made because the spectral quality of that irradiance varies very little, even from day to day. A spectrophotometer has recently been built [3] using the criteria defined by Kumar and Monteith (1982) which gives direct measurements of the ratio of infra-red (780-940 nm) to red (600-660 nm) radiation reflected from a surface. Recent research has concentrated on measuring and interpreting the infra-red to red spectral ratios of the light reflected from a canopy (Steven *et al.*, 1983).

The implication of the work of Kumar and Monteith (1982) is that the fraction of light intercepted by a crop can be reliably estimated by remote sensing, and given a knowledge of the efficiency of converting light to dry matter throughout the season, equation 12.1 may be integrated to predict total dry matter yield at harvest,

$$W = \int \epsilon f S \, dt \qquad 12.2$$

This approach was recently tested in a study on sugar beet. The relation between the spectral ratio and *f* was determined on a crop grown near Broom's Barn. On the same crop weekly measurements of dry weight increase were related to the measured amounts of intercepted radiation to calculate a mean value of ϵ. Using a spectral ratio meter mounted in the ADAS aerial survey plane, a nearby commercial field of sugar beet was overflown at regular intervals during its growth, starting soon after emergence when about 1% of the ground was covered by leaves. The only assessment in this field was total dry matter at final harvest. There was good agreement between the measured total dry matter, 14.3 t ha^{-1}, and that predicted using the remotely sensed data, 14.0 t ha^{-1}; giving an error of only 3% in the prediction. The spectral ratio is sensitive to chlorophyll content of the foliage and in the presence of a disease or nutrient deficiency it may therefore sense not just *f* in equation 12.1 but ϵf (Steven *et al.*, 1983). However, much more research is required before the usefulness of this aspect of the technique is evaluated.

Measurement of plant organs

Leaf growth

In the analytical approach to the study of plant growth suggested by Monteith an understanding of the environmental factors influencing leaf growth is of paramount importance, because most of the radiation is intercepted by leaves. Destructive measurements have the disadvantages that the sampling errors are large, much labour is needed and therefore the interval between harvests is relatively long. When studying the influence on growth of variations in environmental factors, the time interval between measurements needs to be in keeping with the periodicity of these variations. By contrast, non-destructive measurements can be repeated on the same plants and thus have smaller inherent sampling errors and can detect significantly smaller increments of growth, making frequent assessments worth while. Some environmental factors, particularly those associated with the aerial environment, fluctuate so rapidly that growth response cannot be analysed when discrete measurements are made. In these conditions continuous measurements of growth provide the only way to assess the relation between environment and a response, as opposed to an integral response.

Because leaf growth in grasses consists predominantly of increase in leaf lamina length it can be conveniently studied using auxanometers. Early mechanical forms of this instrument have been used in laboratory studies since the last century (Pfeffer, 1900) and some have been used in field investigations (Williams and Biddiscombe, 1965), although several disadvantages were encountered. First, very careful attention was needed during use and charts had to be changed at least daily. Second, the calculation of hourly rates of extension from the chart records was very tedious. More recently, the displacement has been measured using electronic transducers, either rotary potentiometers (Kleinendorst and Brouwer, 1970), or linear variable differential transformers (LVDT) (Gallagher *et al.*, 1976; Christ, 1978). In many respects these transducers are ideal as the measuring device because the ouput signal is linearly proportional to the extension of the leaf, is independent of ambient temperature over a wide range from below freezing point to above 70°C and at normal operating voltages the sensitivity is about 0.05 mm, which compares favourably with average rates of leaf extension in the field of about 2 mm h^{-1}. However, because of the sensitivity of the LVDT's the auxanometer design must compensate for any apparent changes in leaf length caused by the thermal expansion and contraction of the supporting stand.

In use the LVDT is connected to the tip of a leaf by a small clip and to avoid injury to the leaf or plant the tension is adjusted by a series of counterweights. It is advisable to disregard measurements for about an hour after attachment to allow for slight stretching. As the leaf extends the armature of the LVDT is displaced and the changing output signal can be displayed directly or the rate of extension calculated and displayed using modern microprocessors (Saffell *et al.*, 1979).

It is only necessary to reset the LVDT every few days because an extension range of 100 to 250 mm can be accommodated and these occasions provide an opportunity of checking leaf extension against rule measurements. An advantage of this type of instrument is that the extension of the lamina and sheath sections of a leaf can be distinguished by attaching one to the leaf tip and one to the sheath. A similar approach can be used on leaves of other shapes to measure lamina growth in both width and length (Milford, pers. comm.). The supporting stand should allow easy adjustment of the auxanometer, should not disturb the environment of the plant being measured and should be robust. Monteith *et al.* (1981) recommend a sturdy camera-tripod because it has the additional advantage of being easily moved from plant to plant within a large area.

The major environmental variables controlling leaf extension are temperature and water stress. Fig. 12.5 shows the linear relation between hourly extension rate and meristem temperature for winter wheat measured during two days in April 1975. The measurements during the day and at night fall on the same linear response line and show that for winter wheat leaf extension does not stop at 5°C. Indeed, one advantage of the measurements is that a base temperature for leaf extension can be reliably determined.

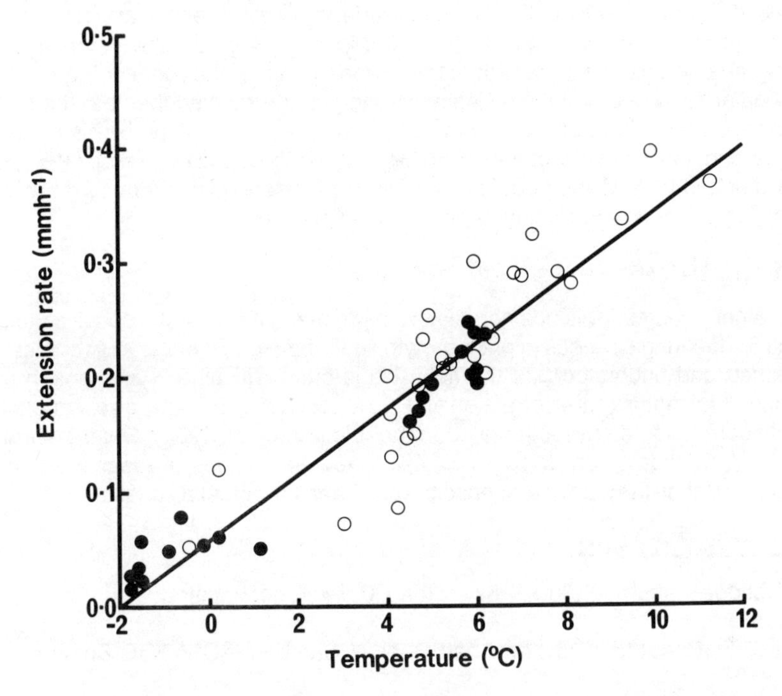

Figure 12.5 The relation between temperature and leaf extension rate of winter wheat measured with an electronic auxanometer. After Gallagher *et al.*, (1979).

Root growth

Compared with the shoots of plants, the root system has received relatively little attention from "whole plant physiologists". This is because the root system is inaccessible, filamentous and delicate. Also roots exist in an environment which is subject to less violent fluctuations than the shoot. Relatively little space will be devoted to roots here, not because they are considered unimportant, but because techniques for examining root systems were recently reviewed by Bohm (1979). Since that review was written most of technical developments which have taken place concern the observation of roots growing alongside transparent glass tubes into the soil. These tubes have the advantage of allowing non-destructive observations, which require relatively little labour and time, and overcome the disadvantage that spatial variation is confounded with variation in time wherever measurements cannot be repeated at any particular location. Studies examining the relation between observations of rooting alongside glass tubes with roots washed from soil cores show that the former method underestimates root density within approximately 200 mm of the soil surface and overestimates the density at depth (Gregory, 1979; Bragg *et al.*, 1982). Overestimates at depth probably indicate that roots are preferentially growing along the glass/soil interface, and recent work at Letcombe Laboratories with tubes inserted into the soil at non-vertical angles should help to overcome this problem (Bragg *et al.*, 1982). However, augering and inserting the longer tubes that will be required to achieve the same depth below the soil surface may be technically difficult. The underestimates of root density near the soil surface by the glass tube technique is less understandable unless physical damage is done to either the roots or the soil when the tubes are inserted or the physical environment around the tube near the surface is atypical of the bulk soil and unsuitable for root growth. These problems need, and are receiving, further investigation. Both Gregory (1979) and Bragg *et al.* (1982) have found much-improved methods for making the observations from tubes with an endoscope or closed circuit television camera respectively.

Conclusions

Unlike other chapters, this has not dealt with the measurement of a specific variable, but with measuring changes in plant growth so that the effect of the environment can be defined and understood. In the field this is often difficult as variations in many environmental features are correlated and the responses of plants are so complex that multiple correlation analyses have been the rule (Monteith, 1981). Our intention has been to show that it is now possible, using new techniques of measurement and analysis, to define the plant's response to specific environmental variables.

Manufacturers

1. Cambridge Instruments Ltd., Rustat Road, Cambridge, England.

2. Delta-T Devices Ltd., 128, Low Road, Burwell, Cambridge CB5 0QE, England.

3. Macam Photometrics Ltd., 10 Kelvin Square, Livingstone EHS4 5DG, Scotland.

References

Anderson, M.C. (1971). Radiation and crop structure. *In* Plant Photosynthetic Production: Manual of Methods, eds. Z. Sestak, J. Catsky and P.G. Jarvis, pp. 412-466. The Hague: Dr W. Junk.

Blackman, V.H. (1919). The compound interest law and plant growth. Ann. Bot., **33**, 353-360.

Böhm, W. (1979). Methods of Studying Root Systems, Ecological Studies 33, Berlin: Springer-Verlag.

Bragg, P.L., Henderson, F.K.G. & Govi, G. (1982). A method for studying growth of cereal roots in the field. Letcombe Laboratory Ann. Report. 1981, pp.21-22.

Briggs, G.E., Kidd, R. & West, C. (1920). Quantitative analysis of plant growth. Ann. appl. Biol., **7**, 103-123, 202-223.

Causton, D.R. & Venus, Jill C. (1981). The biometry of plant growth. London: Edward Arnold.

Christ, R.A. (1978). The elongation rate of wheat leaves. J. exp. Bot., **29**, 603-610.

Cipra, J.E., Noguerapena, N.E., Bryson, M.C. & Lueking, M.A. (1980). Forage production estimates for irrigated meadows from Landsat data. Agron. J., **72**, 793-796.

Evans, G.C. (1972). The Quantitative Analysis of Plant Growth. Oxford: Blackwell Scientific Publications.

Fisher, C. (1971). The new Quantimet 720. Microscope, **19**, 1-20.

Gallagher, J.N., Biscoe, P.V. & Saffell, R.A. (1976). A sensitive auxanometer for field use. J. exp. Bot., **27**, 704-716.

Gallagher, J.N., Biscoe, P.V. & Wallace, J.S. (1979). Field studies of cereal leaf growth. J. exp. Bot., **30**, 657-668.

Gregory, P.J. (1979). A periscope method for observing root growth and distribution in field soil. J. exp. Bot., **30**, 205-214.

Henicke, D.R. (1963). Note on estimation of leaf area and leaf distribution in fruit trees. Can. J. Pl. Sci., **43**, 597-598.

Hudson, H.G. (1939). Population studies with wheat. I. Sampling. J. agric. Sci., Camb., **29**, 76-110.

Hunt, R. (1982). Plant Growth Curves. London: Edward Arnold.

Jaggard, K.W., Lawrence, D.K. & Biscoe, P.V. (1982). An understanding of crop physiology in assessing a plant growth regulator on sugar beet. *In* Chemical Manipulation of Crop Growth and Development, ed. J.S. McLaren, pp.139-150. London: Butterworths.

Kleinendorst, A. & Brouwer, R. (1970). The effect of temperature of the root medium and of the growing point of the shoot on growth, water content and sugar content of maize leaves. Neth. J. agric. Sci., **18**, 140-148.

Kumar, M. (1981). Spectral reflectance and light interception by crop canopies. Ph.D. thesis, University of Nottingham.

Kumar, M. & Monteith, J.L. (1982). Remote sensing of crop growth. *In* Plants and the Daylight Spectrum, ed. H. Smith. London: Academic Press.

Kvet, J. & Marshall, J.K. (1971). Assessment of leaf area and other assimilating plant surfaces. *In* Plant Photosynthetic Production: Manual of Methods, ed. Z. Sestak, J. Catsky and P.G. Jarvis, pp.517-555. The Hague: Dr W. Junk.

Kvet, J., Ondok, P.J., Necas, J. & Jarvis, P.G. (1971). Methods of growth analysis. *In* Plant Photosynthetic Production: Manual of Methods, ed. Z. Sestak, J. Catsky and P.G. Jarvis, pp.343-391. The Hague: Dr W. Junk.

Lawes, J.B. (1850). Amount of water given off by plants. J. hort. Soc. Lond., **5**, 3-28.

Lawrence, J.T. (1969). Area measurement of irregular shapes. Visual, **7**, 10-15.

McNicol, J.W., Ng, S.C.M. & Kidger, R. (1982). A data capture system for agricultural research based on a microcomputer. Exp. Agric., **18**, 255-265.

Monteith, J.L. (1977). Climate and the efficiency of crop production in Britain. Phil. Trans. R. Soc. Lond. Ser. B., **281**, 277-294.

Monteith, J.L. (1981). Climatic variation and the growth of crops. Q. Jl. R. met. Soc., **107**, 749-774.

Monteith, J.L., Gregory, P.J., Marshall, B., Ong, C.K., Saffell, R.A. & Squire, G.R. (1981). Physical measurements in crop physiology. I. Growth and Gas Exchange. Exp. Agric., **17**, 113-126.

Pfeffer, W.F.P. (1900). The physiology of plants. 2nd Ed. Trans. Ed. A.J. Ewart, Oxford.

Saffell, R.A., Biscoe, P.V. & Gallagher, J.N. (1979). A differentiator for use with an electronic auxanometer. J. exp. Bot., **30**, 199-204.

Steven, M.D., Biscoe, P.V. & Jaggard, K.W. (1983). Estimation of sugar beet productivity from reflection in the red and infra-red spectral bands. Int. J. Remote Sensing, **4**, 325-334.

Szeicz, G. (1974). Solar radiation in crop canopies. J. appl. Ecol., **11**, 1117-1156.

Watson, D.J. (1952). The physiological basis of variation in yield. Adv. Agron., **4**, 101-145.

Watson, D.J. (1958). The dependence of net assimilation rate on leaf area index. Ann. Bot., N.S. **22**, 37-54.

Wiegand, C.L., Richardson, A.J. & Kanemasu, E.T. (1979). Leaf area index estimates for wheat from Landsat and their implications for evapotranspiration and crop modelling. Agron. J., **71**, 336-342.

Williams, C.N. & Biddiscombe, E.F. (1965). Extension growth of grass tillers in the field. Aust. J. agric. Res., **16**, 14-22.

Chapter 13: Instruments and instrumenters

L.D. Incoll
Department of Plant Sciences,
University of Leeds, Leeds LS2 9JT, England.

Instruments

In the preceding chapters authors have described how instrumentation for studying the environmental physiology of plants in particular has advanced since 1971. If advances appear slight in some areas then it is either because instruments were already highly developed e.g. thermopile radiometers, because there were particular difficulties in developing suitable instruments e.g. for measuring environmental variables in soil, or because possible new instruments had a limited potential market e.g. instruments for measuring small differences in concentration of oxygen against large backgrounds. On the other hand there have been significant improvements of old instruments and developments of new instruments which measure environmental variables or the performance of plants. The prospects for the future are exciting.

Some of the new instruments have been developed for or by plant scientists because they have been so restricted in their application e.g. the *in situ* thermocouple psychrometer (Campbell, Chapter 11) was built specifically for measuring water potential of leaves.

On the other hand some devices could well see wider application outside plant sciences if they were better known. As Jarvis and Sandford (Chapter 3) have noted earlier, few manufacturers have adopted the split-tube method of calibrating infra-red gas analysers differentially. The fabrication of small critical flow nozzles (Parkinson and Day, 1979) makes possible the simple control of flow rate of gases in the field and, if compressed gases are used, without the need for a source of electrical power. Parkinson and Day have built sets of nozzles of differing sizes into instruments to make a new class of gas-mixing devices. They have used the effect of temperature on the equilibrium between pairs of salt hydrates to produce an instrument which generates gas streams of various water vapour pressures at the flick of a switch (Parkinson and Day, 1981).

The major advance in the measurement of radiation has been the acceptance of the proposition, championed by McCree, that photometric measurements of photosynthetically-active radiation are inappropriate and unacceptable in plant science. Fortunately the prophecy of Downs and Hellmers (1975) that

"We are likely to continue to use the illumination unit (lux), despite irrelevance and high errors of measurement, because instrumentation is relatively inexpensive and because a large number of investigators have illumination meters."

has not been fulfilled. The abandoning of the luxmeter and foot-candle meter was hastened by the availability of reasonably-priced sensors calibrated to measure the flux of photons in appropriate wavebands and traceable to good international standards. On the other hand measurements of photomorphogenetically-active radiation environment have been confused by the use of inappropriate instruments for many years, for example photometric sensors with maximum sensitivity in the green were used and much of those data must be almost useless. There are now two commercial instruments (Macam [3], Skye [4]) that measure the far-red to red ratio, so that, for modest expenditure, every laboratory could have access to one.

Of course environmental physiologists benefit from general developments in the science of measurement. One obvious example since 1971 has been the Vaisala sensor which was developed for measuring relative humidity (h) during the rapid ascent of a radiosonde. In earlier chapters both Day and Parkinson have pointed to the significance of this sensor in measuring h and for incorporation into the class of porometers that measures transpiration from plant surfaces. Long has drawn our attention to the inadequate measurement of flow that prevailed in 1971. As Monteith pointed out in his introduction the pursuit of precision in gas analysis is questionable if another variable such as flow is measured with very low precision. Mass flow meters are used increasingly in plant sciences; their technology has advanced so that meters are available where attitude is not critical and where the range of measurement can be changed readily.

Rapid development of technology based on silicon has given us first miniature integrated circuits and then microprocessors which have been incorporated into instruments of increasing versatility and reliability. This same technology is now being applied to the development of better sensors to supply reliable data to microprocessors. So far the performance of sensors per unit price has lagged behind that of integrated circuits. Demand for cheaper sensors will come from the defence industry, the automobile industry, the domestic appliance industry and, in biology, from medical applications. Costs of development will be such that sensors will only be produced for such large markets. If silicon-based technology is adapted to producing sensors that respond to all the energy domains – radiant, mechanical, thermal, electrical, magnetic and chemical – then the excellent facility of that technology to mass produce chips with great accuracy should give us a versatile range of sensors that are small, cheap and thoroughly reliable. The technology will enable integrated sensors with inbuilt intelligence to be constructed. Non-linearity of response will not be a problem because a look-up table or an equation relating output to the actual value of the measurand could be held in memory. These will be set up by the manufacturer but could be altered later by serial or parallel input signals if the sensor ages. Sensitivity to temperature could be compensated by including a temperature sensor within the integrated sensor. With built-in processing, output could be in serial digital form on external demand rather than as a continuous signal. Diagnostic routines and alarms could be included.

Silicon sensors for the chemical energy domain are of particular interest to environmental physiologists. A prototype, miniature Clark cell has been made for measurement of partial pressure of O_2. The semi-conducting properties of silicon are not used, except for an integral temperature sensor. The silicon simply acts as a substrate for deposition of electrodes, a polymer matrix holding the electrolyte and the semi-permeable membrane (see Engels and Kuypers (1983) for details – in the diamond jubilee issue of J. Phys., E. (Scientific Instruments) which is devoted to "Sensors and their Applications"). Unfortunately miniaturisation means that the life of the electrode is limited by the amount of electrolyte to a few days. Electrochemical cells can also be made where silicon acts as substrate or where the exposed gate of a field effect transistor (FET) is covered with an ion-selective membrane (to make an ISFET) so that the source-drain current depends on the concentration of the selected ion. Though ISFETs sensitive to H^+, Ca^{++}, CO_3^{--} and NO_3^- have been made they are not yet in commercial production.

A recent survey of future needs for sensors in agricultural research and development and in agriculture (ARC, 1982) revealed firstly that as there was a commitment to more extensive sampling of environments, particularly of the soil, and of populations of plants, cheaper sensors were required. The provision of improved sensors for the variables temperature and humidity was accorded the highest priority. Then followed the variables: concentration of ions, partial pressures of O_2 and CO_2 and water potential for which new sensors were needed particularly for measurements in soils and near roots but also in tissues. It will be interesting to see how soon these needs are met by new silicon sensors of the kind described above being developed for other industries.

Optics will also play an increasing part first in measurement, through the use of lasers as remote sensors (Sheehy, Chapter 2) and through the development of new classes of sensors which exploit the properties of optical fibres and second in control, through the use of optical actuators which will enable microcomputers to directly control high power A.C. or D.C. devices. Fibre optic sensors are being developed where the measurand modulates the light emerging from the fibre by changing its intensity, wavelength or polarisation e.g. in fluorescence probes. The modulated light returns by second fibre. Alternatively the properties of the fibre itself may be changed by the measurand e.g. when a fibre is stretched by displacement the phase of the light in the fibre is altered. Optical fibre links are free of electromagnetic interference and optically-based instruments can be operated over long distances without electrical connection between the sensor and the instrument. The scope of future development is great (see Culshaw, 1983). When multiplexing systems are developed a number of signals could be taken from a remote site along one fibre. Miniature fibre sensors will enable measurement in relatively inaccessible places e.g. within a sward. Integrated optical systems fabricated on silicon chips will enable remote processing of signals before transmission.

Instrumenters

Instruments are only the tools of science. The value of good quality instruments to the acquisition of knowledge and solving of problems also depends as well on the quality of the instrumenters. That quality is evident in the care and rigour with which they use instruments and in the clarity, lack of ambiguity, and precision with which they report their results. I will now focus on aspects of the practise of using instruments and of reporting results obtained with them, aspects which have been touched on in earlier chapters and in which we might hope that there will be improvements in the immediate future.

Terminology

Instruments are used to measure physical quantities. Since 1971 there have been a number of serious attempts by groups of scientists, particularly in the USA, to clarify and unify the terminology of their own subset of physical quantities e.g. the Crop Science Society of America (Shibles, 1976), the American Society for Horticultural Science (ASHS, 1977), the North Central Regional Growth Chamber Committee of the U.S. Department of Agriculture-Science Education Agency (Tibbitts and Kozlowski, 1979), and the American Society of Agricultural Engineers (ASAE, 1982). Such positive collective action is admirable and unusual. I recommend the last two references to all scientists who use plant growth chambers.

Fluence rate. Probably the most vexed area at present is that of the terminology of physical quantities for radiation and particularly for the waveband 380 to 780 nm. In an earlier chapter Sheehy has referred to the paper of Bell and Rose (1981) on that subject and drawn attention to the great diversity of terms in current use. The major problem in my opinion is the increasingly inappropriate use of the term fluence rate to the exent that it is now being recommended by editors e.g. in the instructions to authors in Photochemistry and Photobiology and in Planta, and by authors e.g. McClaren (1978) and Mohr and Schafer (1979).

One of the best treatments of measurement of radiation is that edited by Nicodemus for the U.S. National Bureau of Standards (NBS, 1978). It distinguishes between the physical quantities that are commonly confused in plant science viz. radiant fluence rate, radiant flux density and irradiance and their quantum equivalents, photon fluence rate, photon flux density and incident photon flux density. Flux density is the net flux per unit area of energy or photons through a surface of fixed orientation, from (or to) all (or any) directions within a hemisphere on one side of it. Thus it is a directed-surface distribution of radiation. As Bell and Rose (1981) note, it is not easy to measure because absorptance, emittance and transmittance of the surface must be known before the net flux can be calculated from the incident absorbed, emitted and transmitted fluxes. It is the appropriate quantity to relate to a reference surface that intersects a beam e.g. a leaf.

Thus $$W(x,y) = d\Phi(x,y)/dA \qquad 13.1$$

where $W(x,y)$ is the first flux density at a point (x,y). $d\Phi$ is the flux or power through a surface element of area dA containing the point (x,y).

Its units are mol m^{-2} s^{-1} or q m^{-2} s^{-1} for flux density of photons (q being number of particles) and W m^{-2} (J m^{-2} s^{-1}) for flux density of energy. It is more usual to measure the incident flux density which for radiant energy is called the irradiance. (There is no special name for incident photon flux density although Bell and Rose (1981) have suggested photon irradiance and Woodward and Sheehy (1983) have suggested quantum irradiance.) These quantities are measured by planar detectors. The naming of this quantity may seem strange to those who associate density with the quotient of a quantity by volume e.g. mass density, but density is often added to the name of quantity for a flux to show the quotient of the quantity by a surface e.g. electric current density and magnetic flux density (ISO, 1981).

By contrast fluence rate is the flux of energy or photons per unit cross-sectional area incident on a spherical volume element.

Hence $F_t(x,y,z)=\mathrm{d}\Phi(x,y,z)/\mathrm{d}a$ 13.2

where $F_t(x,y,z)$ is the fluence rate at point (x,y,z) and $\mathrm{d}\Phi(x,y,z)$ is the flux or power incident on a spherical volume element of cross-sectional area $\mathrm{d}a$ centred on the point (x,y,z).

Thus it is an omni-directional-surface distribution. It is measured with a spherical detector that responds equally to incident radiation from all directions from a full sphere surrounding it. Although two companies (LICOR [2], Biospherical Instruments [1]) make spherical detectors, many of the plant scientists using fluence rate are not using such detectors. There is no reference surface and fluence rate has a unique value at each point in space. It is a useful physical quantity in limnology and oceanography. It has various other names e.g. because it is a scalar field quantity its radiant energy form is sometimes called scalar irradiance. The CIE Technical Committee on Photometry and Radiometry in September 1977 adopted the term spherical irradiance. Bell and Rose (1981) point out that this physical quantity is really just field strength. Its units are identical to those for flux density and irradiance. Perhaps it is this last fact that is the source of confusion; it may lead to the assumption that the two physical quantities are identical. It is a mistake to judge a quantity by its units rather than by its formal definition and each recommendation for a physical quantity should be supported by such a definition, a symbol and a unit.

All the published official definitions of fluence rate are for omni-directional distributions of radiation (NBS, 1978; ISO, 1980a, 1980b; ICRU, 1980; ASAE, 1982). Indeed the ASAE Guidelines actually state that "fluence measurements must be taken with spherical sensors and cannot be derived from measurements taken with any plane surface detectors". Despite this concordance of definitions Mohr and Schafer called the "spherical" definition "more precise but biologically largely irrelevant". The definition of a physical quantity cannot be "more" or "less" precise.

While the quantity energy fluence rate occurs in the international standards both for light and related electromagnetic radiations and for nuclear reactions and ionising radiations (ISO 1980a, 1980b), the term photon- or more generally particle-fluence rate only appears in the latter international standard with particle flux density given as a synonym. As the Royal Society (1975) has observed "The highest international authority with respect to the names, definitions, and symbols for physical quantities ...... is the ISO". It is a pity therefore that the ISO perpetuates this unfortunate synonomy, although it does not alter my conclusion that fluence rate applies only to omni-directional radiation measured by spherical detectors.

"New" and "old" diffusive conductances. In 1977 Cowan advocated the adoption of the dimensions $\mathrm{N\ L^{-2}\ T^{-1}}$ (SI unit: $\mathrm{mol\ m^{-2}\ s^{-1}}$) for the physical quantity diffusive conductance which to that date had the dimensions $\mathrm{L\ T^{-1}}$. (Thus its reciprocal, diffusive resistance, would have dimensions $\mathrm{L^2\ T\ N^{-1}}$ instead of $\mathrm{T\ L^{-1}}$.) The arguments for such dimensions are persuasive when conductance is derived from measurements with instruments that measure partial pressure or mole fraction of gases and I have no quarrel with their adoption. However it is convention that a single physical quantity should have only one set of dimensions even though one set of dimensions can apply to more than one physical quantity, e.g. $M L^2 T^{-2}$ for energy and torque, so it is unfortunate that Cowan did not use a new name for the new physical quantity.

The name conductance cannot be qualified by the adjective "molar" because by international agreement (IUCAP, 1979) molar is restricted to the meaning "divided by amount of substance" e.g. molar volume which has the unit $m^3\ mol^{-1}$. The choice might be between amount-of-substance conductance by analogy with the naming of amount-of-substance concentration (unit symbol – $mol\ m^{-3}$) or mole conductance by analogy with mole fraction.

The "old" diffusive conductance and the substantial existing literature in which it and its reciprocal diffusive resistance are used are neither invalidated nor made less valuable by the "new" terminology. We will continue to refer our students and our associates to that literature. It is not too much to ask researchers to state at least once in the text of their papers and on figures and tables, that they are using amount-of-substance or mole conductances.

Units

One of the significant changes of the last 12 years has been the adoption of the metric system of measurement as the legal and official system of weights and measures by the English-speaking countries including Australia, New Zealand, Canada, Britain and the United States of America. In 1977 and 1981 Long, Ashmore and I tried to encourage authors and editors to use SI, the current version of metric, for reasons which should not need reiterating. The advantages of a common system of units between branches of science are as evident as those of a common spoken language between people. For me the most compelling reason for adherence to SI is educational – students who have studied biology, physics and chemistry at school have learnt SI. It is arrogant and irresponsible to use outdated units routinely in print but especially in teaching.

In the field of environmental physiology, many of the old problems have gone – we seldom see $mg\ dm^{-2}\ h^{-1}$, ppm or vpm; the einstein is on the way out; I think that we have seen the end of cgs units. Perhaps the most tenacious non-SI unit in our discipline is the bar. In a recent book Turner and Kramer (1980) retained the unit of bars (sic) "because it is understood by a wide range of readers".

It has been suggested earlier in this volume that the system based on the bar gives more memorable numbers of more convenient size conveying a better impression of precision than SI units of pressure. There is no coherent system of units in which all physical quantities can be measured in convenient sizes, particularly when areas and volumes are involved. However, for a range of physical properties where units of pressure are used, the sizes in SI are no less convenient e.g. from a matric potential of air-dry soil of –100 MPa instead of –1000 bar, to the partial pressure of CO_2 in air of +30 Pa instead of +300 μbar , and the same amount of information is present e.g. $-3 \cdot 10^0$ MPa is as precise as $-3\ 10^1$ bar. Furthermore I think that a standard atmosphere of 101.3 kPa is just as memorable as 1.013 bar or 1013 mbar; we probably remember best the one that we learn first.

Recording, calculating and reporting results

With new developments in instruments come new pitfalls and problems. For example many instruments now have digital panel meters. These meters often display a number of variables sequentially and not all of the displayed digits will be significant for all of those variables. Indeed even for an instrument measuring only one variable the discrimination of the display often exceeds the discrimination of the measurement. An awareness of a pitfall such as this would be expected of physical scientists and engineers because the subject of measurement is an accepted part of their education. The subject is also well supported by texts (e.g. Hayward, 1977; Barry, 1978; Taylor, 1982). By contrast there is a tendency for biologists to be impatient with colleagues who suggest that biology students be taught the essentials of the science of measurement. They insist that there is already too much biology to teach despite the fact that good quantitative measurements are the foundation of modern biology. The problem is exacerbated first by the requirement of many universities that potential biology students study biology and chemistry in the last one or two years of secondary school, which usually means instead of physics or mathematics, and secondly by the shortage, at least in the United Kingdom, of well-qualified teachers of physics and mathematics. In attitude and experience many undergraduate biologists therefore enter university inadequately prepared for quantitative biology. Until these problems at pre-university level are sorted out, the situation is only likely to be rectified by changes in the emphasis of teaching of undergraduates.

One of the most important developments in instruments for handling data has been that of the microprocessor. Its potential for measurement and control either as part of an instrument or as the central processing unit of a microcomputer has yet to be fully realised. There are traps for the inexperienced and anyone contemplating purchasing a "hobby" microcomputer for serious scientific work or programming microcomputers in an unstructured language like BASIC should regard Pinches' cautionary chapter (no. 9) as compulsory reading. Finney (1981) has also cautioned biologists about the misuse of computers large and small. Ryder (1983) has drawn attention to the problem common to electronic calculators and microcomputers viz. the restricted number of significant digits which can be handled by their processing registers and which limit the accuracy of an 8-bit processor to $\pm 1\%$. The limited number of significant digits will be important if the microcomputer is being used during the course of an experiment to carry out statistical calculations involving squares because, as large numbers of digits are accumulated, the least significant digits have to be dropped off. Ryder's useful paper describes methods for overcoming these restrictive properties of microcomputers.

I do not want to create the impression that I am pessimistic about the quality of instrumenters; to do so would be arrogant and insulting. However new techniques and new instruments produce new traps and new problems; we need to have them pointed out to us and we need to be reminded of the old ones. We are served well to these ends by the few journals that provide space for critical discussion of techniques, terminology and units, by recent books on instrumentation (e.g. Woodward and Sheehy, 1983) and by the admirable series of papers in Experimental Agriculture called "Methodology of Experimental Agriculture" of which Ryder's paper is an example. With these props and the knowledge of the inventiveness of professional designers of instruments I look forward eagerly to the insights into environmental physiology that advances in instrumentation will bring.

Manufacturers

1. Biospherical Instruments, Inc., 4901 Morena Boulevard, Building 31003, Rose Canyon Business Park, San Diego, California 92117, U.S.A.

2. LI-COR, Inc., P.O. Box 4425, 4421 Superior St., Lincoln, Nebraska 68054, U.S.A.

3. Macam Photometrics Ltd, 10 Kelvin Square, Livingstone, EHS4 5DG, Scotland.

4. Skye Instruments Ltd, The Old Manse, Skeabost Bridge, by Portree, Isle of Skye, Scotland.

References

ARC (1982). Sensors in Agriculture. Report of ARC Working Party on Transducers August 1982. London: Agricultural Research Council.

ASAE (1982). Engineering Practice: ASAE EP411 Guidelines for measuring and reporting environmental parameters for plant experiments in growth chambers. *In* 1982 Agricultural Engineers Handbook (June, 1982) pp.406-409.

ASHS Committee on Growth Chamber Environments (1977). Revised guidelines for reporting studies in controlled environment chambers. HortScience, **12**, 309-310.

Barry, B.A. (1977). Errors in practical measurement in science, engineering and technology. New York: Wiley.

Bell, C.J. & Rose, D.A. (1981). Light measurement and the terminology of flow. Plant, Cell & Environ., **4**,.89-96.

Cowan, I.R. (1977). Stomatal behaviour and environment. Adv. Bot. Res., **4**, 117-228.

Culshaw, B. (1983). Optical systems and sensors for measurement and control. J. Phys., E. (Scientific Instruments), **16**, 978-986.

Downs, R.J. & Hellmers, H. (1975). Environment and the experimental control of plant growth. London: Academic Press.

Engels, J.M.L. & Kuypers, M.H. (1983). Medical applications of silicon sensors. J. Phys., E. (Scientific Instruments), **16**, 987-994.

Finney, D.J. (1981). The misuse of mathematicians, statisticians and computers in agricultural research. Exp. Agric., **17**, 345-353.

Hayward, A.T.J. (1977). Repeatability and accuracy. London: Mechanical Engineering Publications.

ICRU Report 33 (1980). Radiation. Quantities and Units. Washington, D.C.: International Commission on Radiation Units and Measurements.

Incoll, L.D., Long, S.P. & Ashmore, M.R. (1977). SI units in publications in plant science. Commentaries in Plant Science. no. 28, April 1977, *In* Curr. Adv. Plant Sci., pp.331-343.

Incoll, L.D., Long, S.P. & Ashmore M.R. (1981). SI units in publications in plant science. *In* Commentaries in Plant Science, Volume 2, ed. H. Smith, pp.83-96. Oxford: Pergamon.

ISO (1980a). International Standard ISO 31/6-1980 (E). Quantities and units of light and related electromagnetic radiations. 2nd Edition. International Organisation for Standardisation.

ISO (1980b). International Standard ISO 31/10-1980(E). Quantities and units of nuclear reactions and ionising radiations. 2nd Edition. International Organisation for Standardisation.

ISO (1981). International Standard ISO 31/0-1981(E). General principles concerning quantities, units and symbols. 2nd Edition. International Organisation for Standardisation.

IUPAC (1979). Manual of symbols and terminology for physicochemical quantities and units. 2nd Revision. International Union of Pure and Applied Chemistry. Division of Physical Chemistry, Commission on Physicochemical Symbols, Terminology and Units. Oxford: Pergamon.

McClaren, J.S. (1980). The expression of light measurements in relation to crop research. *In* Seed Production. 28th Easter School in Agricultural Science, Nottingham, 1978. ed. P.D. Hebblethwaite, pp.663-670.

Mohr, H. & Schafer, E. (1979). Uniform terminology for radiation: a critical comment. Photochem. Photobiol., **29**, 1061-1062.

NBS (1978). U.S. Department of Commerce, National Bureau of Standards Technical Note 910-2. Self-study manual of optical radiation measurement. Part 1 – Concepts, Chapters 4 and 5. ed. F.E. Nicodemus.

Parkinson, K.J. & Day, W. (1979). The use of orifices to control the flow rate of gases. J. appl. Ecol., **16**, 623-632.

Parkinson, K.J. & Day, W. (1981). Water vapour calibration using salt hydrate transitions. J. exp. Bot., **32**, 411-418.

Shibles, R. (1976). Committee Report: Terminology pertaining to photosynthesis. Crop Sci., **16**, 437-439.

Royal Society (1975). Quantities, Units and Symbols. A report of the Symbols Committee of the Royal Society, London. 2nd Edition.

Ryder, K. (1983). Calculators – how to get the answers right. Exp. Agric., **19**, 15-21.

Taylor, J.R. (1982). An introduction to error analysis: The study of uncertainties in physical measurements. Mill Valley: University Science Books.

Tibbitts, T.W. & Kozlowski, T.T. (1979). Controlled Environment Guidelines for Plant Research. New York: Academic Press.

Turner, N.C. & Kramer, P.J. (1980). Adaptation of plants to water and high temperature stress. New York: Wiley.

Woodward, F.I. & Sheehy, J.E. (1983). Principles and Measurements in Environmental Biology. London: Butterworths.

Index

SOCIETY FOR EXPERIMENTAL BIOLOGY SEMINAR SERIES

A series of multi-author volumes developed from seminars held by the Society for Experimental Biology. Each volume serves not only as an introductory review of a specific topic, but also introduces the reader to experimental evidence to support the theories and principles discussed, and points the way to new research.

1. Effects of air pollution on plants Edited by T.A. MANSFIELD
2. Effects of pollutants on aquatic organisms Edited by A.P.M. LOCKWOOD
3. Analytical and quantitative methods Edited by J.A. MEEK and H.Y. ELDER
4. Isolation of plant growth substances Edited by J.R. HILLMAN
5. Aspects of animal movement Edited by H.Y. ELDER and E.R. TRUEMAN
6. Neurones without impulses: their significance for vertebrate and invertebrate systems Edited by A. ROBERTS and B.M.H. BUSH
7. Development and specialisation of skeletal muscle Edited by D.F. GOLDSPI
8. Stomatal physiology Edited by P.G. JARVIS and T.A. MANSFIELD
9. Brain mechanisms of behaviour in lower vertebrates Edited by P.R. LAMING
10. The cell cycle Edited by P.C.L. JOHN
11. Effects of disease on the physiology of the growing plant Edited by P.G. AYRES
12. Biology of the chemotactic response Edited by J.M. LACKIE and P.C. WILLIAMSON
13. Animal migration Edited by D.J. AIDLEY
14. Biological timekeeping Edited by J. BRADY
15. The nucleolus Edited by E.G. JORDAN and C.A. CULLIS
16. Gills Edited by D.F. HOULIHAN, J.C. RANKIN and T.J. SHUTTLEWORTH
17. Cellular acclimatisation to environmental change Edited by A.R. COSSINS and P. SHETERLINE
18. Plant biotechnology Edited by S.H. MANTELL and H. SMITH
19. Storage carbohydrates in vascular plants Edited by D.H. LEWIS
20. The physiology and biochemistry of plant respiration Edited by J.M. PALM
21. Chloroplast biogenesis Edited by R.J. ELLIS
22. Instrumentation for environmental physiology Edited by B. MARSHALL and F.I. WOODWARD
23. The biosynthesis and metabolism of plant hormones Edited by A. CROZIER a J.R. HILLMAN
24. Coordination of motor behaviour Edited by B.M.H. BUSH and F. CLARAC
25. Cell ageing and cell death Edited by I. DAVIES and D.C. SIGEE
26. The cell division cycle in plants Edited by J.A. BRYANT and D. FRANCIS
27. Control of leaf growth Edited by N.R. BAKER, W.J. DAVIES and C. ONG
28. Biochemistry of plant cell walls Edited by C.T. BRETT and J.R. HILLMAN